T0134905

Springer Theses

Recognizing Outstanding Ph.D. Research

Aims and Scope

The series "Springer Theses" brings together a selection of the very best Ph.D. theses from around the world and across the physical sciences. Nominated and endorsed by two recognized specialists, each published volume has been selected for its scientific excellence and the high impact of its contents for the pertinent field of research. For greater accessibility to non-specialists, the published versions include an extended introduction, as well as a foreword by the student's supervisor explaining the special relevance of the work for the field. As a whole, the series will provide a valuable resource both for newcomers to the research fields described, and for other scientists seeking detailed background information on special questions. Finally, it provides an accredited documentation of the valuable contributions made by today's younger generation of scientists.

Theses are accepted into the series by invited nomination only and must fulfill all of the following criteria

- They must be written in good English.
- The topic should fall within the confines of Chemistry, Physics, Earth Sciences, Engineering and related interdisciplinary fields such as Materials, Nanoscience, Chemical Engineering, Complex Systems and Biophysics.
- The work reported in the thesis must represent a significant scientific advance.
- If the thesis includes previously published material, permission to reproduce this must be gained from the respective copyright holder.
- They must have been examined and passed during the 12 months prior to nomination.
- Each thesis should include a foreword by the supervisor outlining the significance of its content.
- The theses should have a clearly defined structure including an introduction accessible to scientists not expert in that particular field.

More information about this series at http://www.springer.com/series/8790

Yauhen Sachkou

Probing Two-Dimensional Quantum Fluids with Cavity Optomechanics

Doctoral Thesis accepted by
The University of Queensland, Brisbane,
Australia

Author
Dr. Yauhen Sachkou
School of Mathematics and Physics/
ARC Centre of Excellence for Engineered
Quantum Systems
University of Queensland
Brisbane, QLD, Australia

Supervisor
Prof. Warwick Bowen
School of Mathematics and Physics/
ARC Centre of Excellence for Engineered
Quantum Systems
University of Queensland
Brisbane, QLD, Australia

ISSN 2190-5053 ISSN 2190-5061 (electronic)
Springer Theses
ISBN 978-3-030-52768-6 ISBN 978-3-030-52766-2 (eBook)
https://doi.org/10.1007/978-3-030-52766-2

This Springer imprint is published by the registered company Springer Nature Switzerland AG
The registered company address is: Gewerbestrasse 11, 6330 Cham, Switzerland

To my parents, grandparents, Emily.

Per Aspera ad Astra

Supervisor's Foreword

Superfluid helium is a quantum liquid that exhibits a range of counterintuitive phenomena such as frictionless flow. Quantized vortices are a particularly important feature of superfluid helium, and indeed all superfluids, which are characterized by a circulation that can only take prescribed integer values. The theory that recognized their importance in two-dimensional helium was awarded a share of the 2016 Nobel Prize in Physics. However, the strong interactions between atoms in superfluid helium prohibit quantitative theory. Experiments have similarly not been able to observe coherent vortex dynamics.

Yauhen Sachkou's thesis work resolves this challenge, bringing microphotonic techniques to bear on two-dimensional superfluid helium, observing coherent vortex dynamics for the first time, and achieving this on a silicon chip. Until this work, it was thought that strong interactions with the surface would preclude coherent dynamics. The thesis shows that this is not the case, thus allowing the observation of complex vortex dynamics as the superfluid evolves over time, as well as measurements of the properties of superfluid helium in new regimes.

The thesis provides a new technological approach to explore the rich behavior of two-dimensional quantum liquids, including both fundamental questions such as how they evolve and how they dissipate energy, alongside more practical issues such as how their matter–wave character can be used to build precise sensors and other quantum technologies. Some of the most basic questions about quantum vortices still remain open and could be addressed by the approach pioneered here; for example, whether quantized vortices have inertia, or whether their dynamics is affected by a background of normal fluid. Indeed, the technology may even allow new insights into the dynamics of astrophysical superfluids, such as those thought to exist in the core of neutron stars.

I see this work as the start of a new adventure in the physics of strongly interacting superfluids, and I am excited to see where the technology will take us.

Brisbane, Australia
April 2020

Warwick Bowen

Abstract

Superfluidity is an emergent quantum phenomenon intrinsic to a wide range of condensed matter systems, such as ultracold quantum gases, polaritons, and superfluid helium. It had been long believed that condensation into the superfluid phase in two-dimensional matter is precluded due to thermal fluctuations, which destroy the long-range phase coherence. However, superfluidity was experimentally observed in a variety of two-dimensional physical systems. The key to superfluid topological phase transitions is quantized vortices. These elementary excitations and their interactions with phonons and rotons—other types of elementary excitations—determine the dynamics of two-dimensional quantum fluids. Dynamics of superfluids with weak atom-atom interactions, such as Bose-Einstein condensates in dilute gases, are typically well described within the framework of the Gross-Pitaevskii equation. Moreover, behaviour of weakly interacting quantum fluids is subject to an exquisite experimental control enabled by advanced methods of quantum optics. In contrast, a complete microscopic model for superfluids with strong atom-atom interactions, such as superfluid helium, is still an active area of research. This is accompanied by the absence of experimental methods capable of probing thermodynamics and microscopic behaviour of strongly interacting quantum fluids both in real time, and nondestructively in a single shot.

The major goal of the research presented in this thesis is to bring a comprehensive level of control to strongly interacting two-dimensional quantum fluids, such as superfluid helium. We achieve this by leveraging methods of cavity optomechanics, which provides a toolkit for ultraprecise optical readout of superfluid dynamics. The optomechanical system reported in this thesis is comprised of a whispering-gallery-mode optical microcavity coated with a few-nanometer-thick film of superfluid helium. The combination of the small electromagnetic mode volume and the high optical quality factor of these microcavities enables enhanced light-matter interactions at the interface of such a device, allowing the microscopic dynamics of two-dimensional superfluid helium to be probed with unprecedented resolution and precision.

Throughout this thesis, we first describe the theoretical aspects of coupling between light, confined within a high-quality whispering-gallery-mode microcavity, and the mechanical motion of a superfluid helium film. Experimental realization of such an optomechanical system allowed us, for the first time, to track the thermomechanical motion of superfluid helium in real time, *i.e.* faster than the oscillator's decay timescale. Furthermore, by damping and amplifying sound waves on superfluid thin films, we show the ability to control the thermal motion of a quantum fluid.

Exploiting our superfluid optomechanical system, we also demonstrate a new approach to generating strong microphotonic forces on a chip in cryogenic conditions. Utilising the superfluid fountain effect, we are able to control superfluid flow, which is directed towards the localised heat source provided by absorption of light in the material of the optical microcavity. Upon evaporation in the vicinity of the heat source, helium atoms recoil and exert force on the resonator. We experimentally achieve microphotonic forces one order of magnitude larger than is possible with radiation pressure in cryogenic conditions. We utilise these forces to feedback-cool a vibrational mode of the optical microcavity down to 137 mK.

Having demonstrated the ability to track and manipulate thermal excitations in superfluid helium and to control superfluid flow with laser light—prerequisites for generating and tracking quantized vortices in our system—we then move on to study vortex-phonon interactions in two-dimensional superfluid helium. We first develop a theoretical framework for these interactions and rigorously quantify them. This framework allows us to analytically calculate the rate of the vortex-phonon coupling for any arbitrary distribution of vortices within a cylindrical geometry. Moreover, we develop a finite-element method to model the interaction of an arbitrarily distributed vortex ensemble with sound in any arbitrary, perhaps multiply-connected geometry. We furthermore analyse the prospect of detecting a single quantum of circulation in a two-dimensional strongly interacting quantum fluid.

One of the core results of this thesis is the first hitherto experimental observation of the coherent dynamics of nonequilibrium vortex clusters in a strongly interacting two-dimensional superfluid. We achieve this by confining superfluid helium on the atomically-smooth surface of a silicon chip, orders of magnitude more strongly than has previously been possible, resulting in greatly enhanced coherent interactions between vortices. The atomically-smooth surface of the chip combined with a superfluid temperature much lower than the Berezinskii-Kosterlitz-Thouless phase transition temperature enables the vortex diffusivity five orders of magnitude lower than in previous measurements with unpinned vortices in superfluid helium films. The observed microscopic dynamics are supported by point-vortex simulations. Our results open up new prospects for studying two-dimensional quantum turbulence, phase transitions, and dissipation mechanisms in two-dimensional strongly interacting quantum fluids.

Publications related to this Thesis

1. **Y. P. Sachkou**, C. G. Baker, G. I. Harris, O. R. Stockdale, S. Forstner, M. T. Reeves, X. He, D. L. McAuslan, A. S. Bradley, M. J. Davis, W. P. Bowen, Coherent vortex dynamics in a strongly interacting superfluid on a silicon chip, *Science* 366, 1480–1485, 2019.
2. S. Forstner, **Y. Sachkou**, M. Woolley, G. I. Harris, X. He, W. P. Bowen, C. G. Baker, Modelling of vorticity, sound and their interaction in two-dimensional superfluids, *New Journal of Physics* 21, 053029, 2019.
3. G. I. Harris, D. L. McAuslan, E. Sheridan, **Y. Sachkou**, C. Baker, W. P. Bowen, Laser cooling and control of excitations in superfluid helium, *Nature Physics* 12, 788–793, 2016.
4. D. L. McAuslan, G. I. Harris, C. Baker, **Y. Sachkou**, X. He, E. Sheridan, and W. P. Bowen, Microphotonic forces from superfluid flow, *Physical Review X* 6, 021012, 2016.
5. C. G. Baker, G. I. Harris, D. L. McAuslan, **Y. Sachkou**, X. He, W. P. Bowen, Theoretical framework for thin film superfluid optomechanics: towards the quantum regime, *New Journal of Physics* 18, 123025, 2016.

Acknowledgements

Doctor of Philosophy is a challenging and demanding endeavour, which certainly requires a supportive and inspiring environment.

First of all, I would like to wholeheartedly thank my Principal Advisor Warwick Bowen for the exclusive opportunity to come to Australia and to conduct a world-class research on a fascinating subject of superfluidity in two-dimensional helium. Thank you for your guidance, inspiration, patience, enthusiasm, and an excellent example of a great scientific intuition. You made my PhD journey exciting and memorable. Thank you, Warwick!

I have been also incredibly lucky to receive guidance and mentorship from two brilliant scientists and extraordinary people—my Associate Advisors Christopher Baker and Glen Harris. I am forever grateful to you both for your invaluable day-to-day guidance, and for everything you taught me. Chris, thank you for your tremendous support and motivation, especially during the hardships of my PhD journey. Glen, thank you for your kindness and welcoming Australian spirit.

I express my massive gratitude to those who assisted with the proof-reading of this thesis: Christopher Baker, Glen Harris, James Bennett, Emily Mantilla.

I address my immense gratitude to our superfluid team: Christopher Baker, Glen Harris, Eric He (big thank you for the great time spent together in the lab), David McAuslan, Stefan Forstner, Andreas Sawadsky, Yasmine Sfendla. Thank you all for your great passion for superfluid helium!

In our Queensland Quantum Optics Laboratory, I have worked alongside many wonderful people, who have always been a great source of energy and inspiration. I would like to sincerely thank each and every one of you: Erick Romero, James Bennett, Lars Madsen, Christiaan Bekker, George Brawley, Nicolas Mauranyapin, Beibei Li, Varun Prakash, Rachpon Kalra, Muhammad Waleed, Nick Wyatt, Kiran Khosla, Catxere Casacio, Chao Meng, Hamish Greenall, Larnii Booth, Ardalan Armin, Sahar Basiri-Esfahani, Amy van der Hel, Abdullah Al Hafiz, Ulrich Hoff, Yimin Yu, Andreas Næsby, Michael Taylor, Gian-Marco Schnüriger, Leo Sementilli, Fernando Gotardo, Atieh Kermany, Michael Vanner, Victor Valenzuela, Sarah Yu, Shaoli Zhu.

I thank all my office mates, who I have been lucky to share the office with over the past few years: Christopher Baker, Beibei Li, Eric He, Erick Romero, Varun Prakash, Catxere Casacio, Fernando Gotardo, Ulrich Hoff. Thank you for always maintaining our office a lively place!

My deep gratitude is expressed to the people of administration, who have provided an immense assistance throughout my entire PhD candidature: Kaerin Gardner, Murray Kane, Linda Waldrum, Angela Bird, Tara Massingham, Lisa Laletina (Provoroff), Erinn Osmond.

I owe an infinite debt of gratitude to my parents, Pavel and Natalia, for their love, guidance, and tremendous and unconditional support. I also address my deepest and sincerest gratitude to my grandparents.

My very special and immense thank you to my delightful partner Emily. Thank you from the bottom of my heart for your love, kindness, support, inspiration, and patience! Thank you for always being there for me, especially when I need it most.

Last but not least, I would like to thank The University of Queensland, ARC Centre of Excellence for Engineered Quantum Systems (EQUS), and U.S. Army Research Office for their financial support, which made my PhD adventure possible.

Contents

Abbreviations

AC	Alternating Current
AM	Amplitude Modulator
BEC	Bose-Einstein Condensate
BKT	Berezinskii-Kosterlitz-Thouless (phase transition)
CCW	Counter-Clockwise Rotating Wave
CW	Clockwise Rotating Wave
DC	Direct Current
FBS	Fibre Beam Splitter
FEM	Finite-Element Modelling
LO	Local Oscillator
NA	Network Analyzer
PD	Photodetector
PTC	Pulse-Tube Cooler
QOL	Quantum Optics Laboratory
RMS	Root-Mean Square
SA	Spectrum Analyzer
SEM	Scanning Electron Microscope
SMF	Single-Mode Fibre
SNR	Signal-to-Noise Ratio
VDW	Van der Waals
WGM	Whispering-Gallery Mode

Symbols

κ	Quantum of circulation *and* optical decay rate (context dependent)
m_{He}	Mass of a helium atom
α_{vdw}	Van der Waals coefficient
c_3	Third-sound velocity
ϵ_{sf}	Permittivity of superfluid helium
n_{sf}	Refractive index of superfluid helium
a_0	Vortex core size (healing length)
Δf	Third-sound mode splitting
K_{total}	Total kinetic energy
K_{pinned}	Kinetic energy of pinned macroscopic circulation
K_{free}	Kinetic energy of a free-vortex cluster
γ	Dissipation coefficient
T	Temperature
k_{B}	Boltzmann constant
R	Radius
D	Diffusivity
ω_{m} Ω_{m}	Mechanical angular frequency
Γ_{m}	Mechanical decay rate
Q	Quality factor
x_{zpf}	Amplitude of zero-point fluctuations
g_0	Single-photon optomechanical coupling
G	Optomechanical coupling
C	Cooperativity
C_0	Single-photon cooperativity
m_{eff}	Effective mass
Δ	Optical detuning
v_{rms}	Root-mean-square velocity

Chapter 1
Introduction and Overview

1.1 Introduction

Superfluidity is an emergent quantum phenomenon that arises in a number of condensed matter systems at temperatures very close to absolute zero. The first system that was discovered to exhibit superfluidity was cryogenically cooled liquid helium-4 [1, 2]. Below a certain critical temperature helium atoms undergo Bose-Einstein condensation (BEC) into a superfluid state that exhibits strong atom-atom interactions, long-range quantum phase coherence and is described by a macroscopic quantum wave function [3]. The most famous macroscopic manifestation of superfluidity is flow without friction—a phenomenon that closely resembles the resistance-less transport of electrons in superconductors [4]. The elementary excitations existing in superfluids are phonons and rotons—both manifest in the form of sounds waves—and vortices that carry the angular momentum of the fluid and are quantized due to the requirement of continuity of the macroscopic wave function [3]. These elementary excitations and interactions between them determine both the macroscopic thermodynamical properties and microscopic quantum dynamics of superfluids. Thus, the ability to resolve, track, and control the elementary excitations is crucial to the understanding of the nature of superfluidity.

Interactions between elementary excitations are even more important in two-dimensional quantum fluids, where the excitations in one dimension are not thermally activated. It was long believed that, according to the Mermin-Wagner-Hohenberg theorem [5, 6], thermal fluctuations destroy long-range phase coherence, thus precluding Bose-Einstein condensation and, therefore, superfluidity in two-dimensional systems. Nevertheless, superfluidity was found to exist in two-dimensional films of superfluid helium [7]. This apparent contradiction was resolved by Berezinskii, Kosterlitz and Thouless, who realized the crucial importance of quantized vortices in understanding of two-dimensional helium [8–11]. Kosterlitz and Thouless were

Y. Sachkou, *Probing Two-Dimensional Quantum Fluids with Cavity Optomechanics*,
Springer Theses, https://doi.org/10.1007/978-3-030-52766-2_1

awarded the 2016 Nobel Prize in Physics[1] "for theoretical discoveries of topological phase transitions and topological phases of matter" [12, 13]. They discovered that below a critical temperature two-dimensional superfluid helium undergoes a phase transition where initially free vortices bind into pairs. The Berezinskii-Kosterlitz-Thouless (BKT) phase transition is a great example out of a rich range of phenomena that arise due to vortex-vortex and vortex-phonon interactions in two-dimensional quantum fluids. These interactions determine the dynamics of the fluid, giving rise to such phenomena as quantum turbulence [14], dissipation [15], the formation of large-scale Onsager vortices [16], etc. However, while quantized vortices have already been observed in bulk superfluid helium [17] and their dynamics were tracked in real time [18, 19], to date, quantized vortices have never been directly observed in two-dimensional superfluid helium. The evidence of their existence still remains indirect and inferred through measurements of macroscopic thermodynamic properties [7, 20] and splitting of sound modes in two-dimensional helium films [21–23].

Since its original discovery in liquid helium, two-dimensional superfluidity has been also observed in other quantum systems, including ultracold Bose [24] and Fermi gases [25], as well as semiconductor exciton-polaritons [26, 27]. A major advantage allowed by these systems is the ability to control and observe the dynamics of vortices with an exquisite precision using laser light [28, 29]. This has allowed the fascinating properties of superfluids to be explored in new regimes and with an unprecedented level of details. Single quantized vortices have been observed both in two-dimensional BECs of dilute gases [30] and in exciton-polariton systems [26]. Moreover, experiments with both ultracold gases [31] and exciton-polaritons [28] allow the direct imaging of vortex configurations in the superfluid phase. Most recently, quantum turbulence has been observed in these systems showing, for example, the formation of vortex dipoles with negative temperature and large-scale Onsager vortices [32, 33]. However, these experiments are generally limited to the regime of *weak atom-atom interactions*, where the Gross-Pitaevskii equation—also known as a nonlinear Schrödinger equation—provides a microscopic model of the quantum fluid dynamics [34]. Using Feshbach resonances it is possible to reach the strongly interacting regime in ultracold gases [25, 34–39], however, three-body collisions inherent to these systems have so far prevented the observation of vortex dynamics in the strongly interacting regime.

Nowadays, a comprehensive model of the dynamics of strongly interacting superfluids is still lacking. The *strongly interacting regime*, where the atomic scattering length is long compared to the inter-atom separation, is the relevant regime for superfluid helium, as well as for a wide range of other systems from string theory [40, 41] to astrophysics [42], including dark matter [43], the quark-gluon plasma in the early universe [42], and the dense cores of neutron stars [44, 45]. The dynamics of these systems are dominated by interactions between quantized vortices. For instance, the observed glitches in the rotation frequency of neutron stars are thought to result from vortex unpinning events [46]. The lacking of a comprehensive theory of the dynamics

[1] V. Berezinskii passed away in 1980.

of the strongly interacting superfluids motivates experiments to directly probe the microscopic behaviour of these quantum fluids.

1.1.1 Outstanding Questions of Superfluid Dynamics

The vortex dynamics in strongly interacting two-dimensional superfluids are typically predicted through phenomenological point vortex models. However, given the lack of an underpinning microscopic theory, there exists a range of unanswered questions which are subject to contentious debate. For instance, proposed values for the effective mass of a vortex range from zero to infinity [47, 48] and it remains unclear whether vortices should have inertia [47–50]. Moreover, the precise nature of the forces experienced by vortices due to the normal-fluid component (e.g. Iordanskii force) [51–53] is highly controversial. Furthermore, it remains unclear how to treat dissipation given the non-local nature of the vortex flow fields [20, 54].

We will now provide a glance at the questions of the dynamics of two-dimensional strongly interacting quantum fluids which are yet to be answered both theoretically and experimentally:

- direct experimental observation of quantized vortices;
- continuous non-destructive monitoring of the dynamics of single quantized vortices as well as vortex ensembles;
- development of a microscopic model of vortex nucleation and annihilation—crucial for understanding of quantum turbulence [55];
- direct experimental observation of the vortex nucleation and annihilation dynamics;
- observation of Onsager vortices;
- understanding of superfluid dissipation in two dimensions.

The presented list of open questions is by no means exhaustive.

A major advancement in the understanding of the interactions between elementary excitations in two-dimensional superfluid helium was gained through a series of pioneering works by Ellis et al. on measurements of sound modes (third sound—see Sect. 1.10) in confined thin superfluid films [21, 22, 56, 57]. Utilizing an electrical measurement scheme, they observed the interaction of a large number of vortices with third-sound modes on a centimeter-scale circular resonator [21, 22]. Moreover, they modelled this interaction numerically [56].

This thesis builds up on the idea to resolve quantized vortices via their interaction with confined phonon excitations within a thin film of superfluid helium. We leverage the progress of the microfabrication technologies to develop microscale optical resonators, which provide unprecedented confinement to both vortices and phonons co-localized on the surface of the resonator. Improved confinement enables a great enhancement of the vortex-phonon interaction, which is read out optically using the methods of cavity optomechanics. The major goal of this thesis is to establish an exquisite level of control over strongly interacting two-dimensional quantum fluids and to observe superfluid dynamics in real time.

1.2 Thesis Outline

This thesis consists of 6 chapters. A collage reflecting main ideas of the 4 progress chapters is shown in Fig. 1.1.

The remaining part of the current chapter provides the background necessary for understanding the research presented in this thesis. It covers the basics of superfluidity and optomechanics. The building blocks of our superfluid optomechanical system—whispering-gallery-mode optical resonators and third-sound modes in thin films of superfluid helium—are described in the second half of the current chapter.

Chapter 2 introduces a paradigm of cavity optomechanics with thin films of superfluid helium (Fig. 1.1a). It describes the mechanism of the coupling between optical field and the superfluid mechanical motion. This is followed by the presentation of our experimental system. Moreover, the chapter also describes the real-time monitoring and control of the superfluid thermomechanical motion.

Chapter 3 presents our ability to control superfluid flow with light (Fig. 1.1b). It introduces the light-induced superfluid photoconvective forcing of a mechanical oscillator and shows that this force is stronger than its radiation pressure counterpart. The chapter also details how the force from the superfluid flow can be utilized to feedback cool mechanical motion of the oscillator.

Chapter 4 details a theoretical model of the vortex-phonon interactions within a thin film of superfluid helium confined to the surface of an optical microresonator (Fig. 1.1c). Furthermore, it introduces a method for finite-element modelling of the vortex-phonon interactions within an arbitrary domain (not necessarily simply-connected) and for an arbitrary distribution of vortices.

Chapter 5 describes the observation of coherent vortex dynamics in a strongly interacting two-dimensional superfluid (Fig. 1.1d). Moreover, it details the observation of evaporative heating of a vortex cluster. The chapter also introduces a number of models of the vortex dynamics feasible in our system and discriminates between them by comparing to the experimental results.

Chapter 6 summarises the key research outcomes presented in this thesis.

1.3 Quantum Fluids and Superfluidity

A quantum fluid is a condensed matter system which remains fluid (either gas or liquid) even at temperatures close to absolute zero [4]. Namely, at those temperatures, where the effects of quantum mechanics start playing the dominant role, and behaviour of the system can no longer be described with the laws of classical physics. However, a quantum fluid is a system, where not only the effects of

Fig. 1.1 Thesis outline. Each of the subfigures **a–d** reflect the main ideas of the four main progress chapters of the thesis. **a** Chapter 2 introduces a paradigm of cavity optomechanics with thin films of superfluid helium. **b** Chapter 3 presents our ability to control superfluid flow with light. **c** Chapter 4 details a theoretical model of the vortex-phonon interactions in thin films of superfluid helium. **d** Chapter 5 describes the observation of coherent vortex dynamics in a strongly interacting two-dimensional superfluid. Rendering was supplied by C. G. Baker

quantum mechanics are important, but also those of quantum statistics, i.e. characteristic indistinguishability of elementary particles [58, 59]. Two most famous types of quantum fluids are the systems comprised of bosons (superfluid helium-4, Bose ultracold gases), which undergo Bose-Einstein condensation, and the systems comprised of fermions (superfluid helium-3, electrons in some metals), which exhibit a phenomenon of Cooper pairing [58, 59].

The effects of quantum mechanics are significant when the thermal de Broglie wavelength of the particles constituting the fluid becomes comparable with other typical length scales in the fluid, such as interatomic distances [4]. Liquid helium-4 was the first discovered quantum fluid, and the field of low-temperature physics owes its existence to liquid helium [60]. The thermal de Broglie wavelength of helium-4 particles are always greater than the typical interatomic spacing in the fluid, and therefore the effects of quantum mechanics are always important for liquid helium-4 [4].

When cooled below a certain critical temperature, helium-4 atoms undergo Bose-Einstein condensation. Thus, liquid helium-4 undergoes a second-order phase transition and becomes superfluid [3]. The phase of liquid helium below the superfluid phase transition temperature is conventionally called He II, whilst the phase above that temperature—He I. The condensate of helium atoms in the He II phase is described by the macroscopic wave function (also known as "order parameter") [3]

$$\psi(\vec{r}, t) = \psi_0(\vec{r}, t)e^{iS(\vec{r},t)}, \tag{1.1}$$

where $\psi_0(\vec{r}, t)$ is the function amplitude and $S(\vec{r}, t)$ its phase.

The macroscopic velocity $\vec{v}_s$ of the superfluid is the same function of ψ as is the probability current in conventional quantum mechanics:

$$\vec{v}_s = \frac{\hbar}{m_{\mathrm{He}}} \nabla S, \tag{1.2}$$

where m_{He} is the mass of a helium atom. Thus, the superfluid velocity is proportional to the gradient of the phase of the macroscopic wave function ψ of the condensate [3].

1.4 Two-Fluid Model

In the late 1930s three newly discovered properties of He II presented a particular perplexity. Namely, (i) a 'self' filling of a tube only partly immersed in He II; (ii) the thermomechanical effect (elaborated below in Sect. 1.5); and (iii) the viscosity paradox [61]. In 1938, inspired by the idea of F. London that the transition to the He II phase was a manifestation of the Bose-Einstein condensation (BEC) of helium atoms, L. Tisza proposed a two-fluid model [62, 63], which proved to be successful in qualitative explanation of the above baffling experimental observations and was further developed by L. Landau in 1941–1947 who put forward a set of two-fluid equations of motion [64, 65].

The basic idea of the two-fluid model is that He II behaves as if it were comprised of two intermingled fluids, one of which is superfluid and exhibits zero viscosity, while the second one is normal fluid and behaves like a classical viscous liquid [3, 4]. Both of these fluid components have their own independent densities and the total density ρ of liquid helium II is the sum of the densities of the two components [3]:

$$\rho = \rho_s + \rho_n, \tag{1.3}$$

where ρ_s and ρ_n are densities of respectively the superfluid and normal fluid constituents. Both ρ_s and ρ_n depend on temperature, as shown in Fig. 1.2. It was found empirically that at very low temperatures, near $T = 0$, almost all helium atoms condense in BEC, which results in the total density being comprised almost entirely of only the superfluid component, i.e. $\rho_s \sim \rho$ and $\rho_n \sim 0$. The situation is opposite close to the λ-point temperature, where nearly all of the atoms belong to the normal-fluid constituent, i.e. $\rho_s \sim 0$ and $\rho_n \sim \rho$ (Fig. 1.2). At absolute zero temperature all of the helium atoms participate in the superflow, and this particle fraction gradually drops as the temperature is raised [4] (Fig. 1.2).

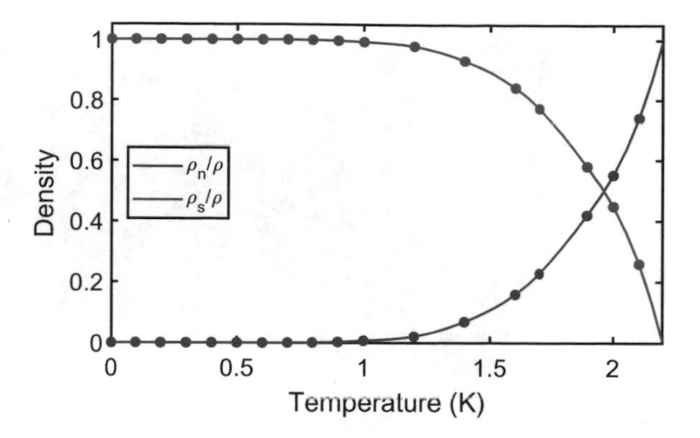

Fig. 1.2 Density of the superfluid and normal-fluid components as a function of temperature below the λ-transition temperature. ρ_s and ρ_n are densities of respectively the superfluid and the normal-fluid components, and $\rho = \rho_s + \rho_n$ is the total fluid density. Dots correspond to the density values taken from Ref. [66] and solid lines are interpolations

The two-fluid hydrodynamics entails some important properties of both fluid constituents. The normal-fluid component carries entropy as a result of the chaotic motions of atoms in the normal fluid. In contrast, the superfluid component is comprised of the Bose-Einstein condensate which embodies a single many-body quantum state, and therefore does not carry entropy. Thus, heat in superfluids is conducted purely by the normal-fluid component.

Another important characteristic of the two-fluid model is that there is no friction between the superfluid and normal-fluid components when they move "through each other", i.e. there is no momentum exchange between them [3, 4]. However, it is important to note that the superfluid and normal-fluid components are inseparable, i.e. upon randomly picking a helium atom from liquid He II it is impossible by any means to tell whether the atom belongs to the superfluid or the normal-fluid component—all helium atoms are identical and all together comprise liquid He II.

In the following, we will be referring to He II as just 'superfluid helium'.

1.5 Superfluid Thermomechanical Effect

One of the consequences of the two-fluid model—namely that only the normal-fluid component carries entropy—is a peculiar thermomechanical effect. It is best illustrated in an experiment where two containers filled with He II are linked via a narrow channel (superleak) [4, 65]. The temperature and pressure in both containers are equal at the beginning of the experiment. A pressure increase in one of the containers results in a flow of helium towards the second one. This results in a thermal imbalance, whereby the temperature in the container with less helium is raised, whereas the temperature in the other is decreased. This is explained by the fact that the viscous normal-fluid component is clamped within the very thin capillary connecting the containers, whilst the superfluid component can freely flow from one container to the other. Since the normal-fluid component is immobilized, it cannot transfer entropy. Hence, there is a mass flow of He II but no heat transfer. And in the

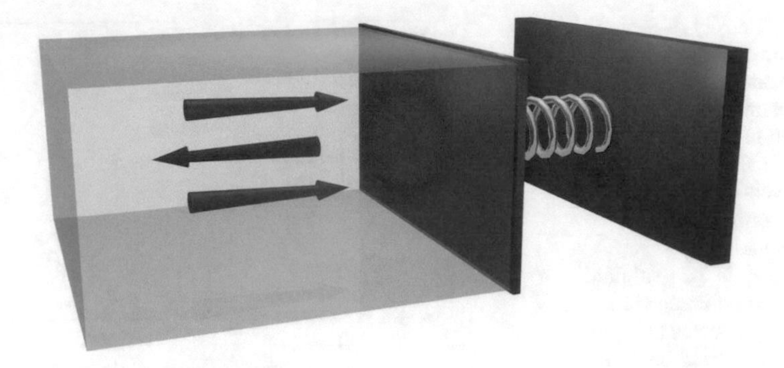

Fig. 1.3 A prototype configuration for exploitation of the superfluid thermomechanical effect in optomechanical systems. In this set-up, a heat source (red glow) is placed onto a mechanically compliant element. This initiates flow of the superfluid component (blue arrows) towards the heat source and counterflow of the normal-fluid component (red arrow) in the opposite direction. Upon arrival at the mechanical element, the superflow imparts momentum onto the oscillator. Rendering was provided by C. G. Baker

container with less helium internal energy is redistributed within a smaller amount of He II, resulting in its temperature increase [4, 65]. Remarkably, making the pressure in the containers equal again brings the system to its initial condition, indicating that the process is reversible [67].

A reversal of the experiment, whereby heating creates a pressure difference, gives rise to the famous *fountain effect*, which was first observed by J. Allen and H. Jones in 1938 [68]. In this experiment, a narrow-necked flask filled with dense powder (superleak) and open at the bottom is immersed in He II. When the flask is heated, the temperature difference generates a net superflow through the superleak in order to minimize the superfluid gradient. The superflow rushes towards the heat source with high velocity and spouts out above the free surface of He II through the narrow neck of the flask.

The characteristic of superfluid helium to flow towards regions with raised temperature is exploited in our experiments presented in this thesis. For example, in Chap. 3 we utilize the thermomechanical effect to apply an optical force to a mechanical oscillator at the microscale. Namely, if a light-induced heat source is placed onto the mechanically compliant element, such as in Fig. 1.3, the superfluid component flows towards the heat source and thereby imparts momentum onto the mechanical oscillator.

1.6 Sound Waves in Superfluids

Collective mechanical excitations propagating in superfluids in the form of waves are conventionally called "sound" [3, 67]. The two-fluid model of superfluid helium gives rise to different types of sound, depending on whether the helium is bulk, two-dimensional, or confined within a fine capillary. Here we provide a brief overview of the commonly distinguished types of sound.

- *First sound* is intrinsic to bulk superfluid helium. It constitutes as fluid density fluctuations driven by changes in pressure at almost constant entropy (i.e. sound velocity only very weakly depends on temperature) [61]. In this case both normal and superfluid components oscillate in phase with each other. First sound is analogous to ordinary sound in a classical fluid, which requires finite compressibility of the system [69].
- *Second sound* manifests as entropy fluctuations driven by changes in temperature at almost constant total density [61, 70, 71]. In this case normal and superfluid components oscillate out-of-phase relative to each other, thus, maintaining zero net flow of mass [3]. Therefore, in the limit of negligible thermal expansion, second sound can be treated as pure temperature oscillations. Similar to the first sound, second sound is also characteristic to bulk superfluid helium. The velocity of second sound strongly depends on temperature and becomes zero at the superfluid phase transition temperature. Thus, this type of wave is specific only to superfluids and does not exist in ordinary classical liquids [65].
- *Third-sound* is a surface wave propagating on a thin film of superfluid helium on a solid substrate. Here, the superfluid component oscillates parallel to the substrate while the normal component is viscously clamped. This type of sound is central to this thesis and will be elaborated in Sect. 1.10.
- *Fourth sound* is a wave propagating along a fine capillary with a diameter smaller than the viscous penetration depth [3, 65]. This means that, similar to the third sound, the normal component is immobile due to viscosity and only the superfluid component can sustain motion. Therefore, the fourth sound is accompanied not only by oscillations of density, but also of pressure, temperature, entropy and the relative superfluid density ρ_s/ρ, where ρ is the total density. Fourth sound propagates with almost no damping [67].

1.7 Quantized Vortices

Vorticity is an ubiquitous phenomenon in nature. It is inherent to a wide range of condensed matter systems from classical fluids [72] to superconductors [3]. Observation of vorticity in a classical fluid is a trivial endeavour: all is required is, for instance, stir water in a glass with a spoon. However, the problem of vorticity is much more delicate in relation to quantum fluids. Unlike their classical counterparts, vortices in quantum fluids possess circulation quantized in terms of the Planck constant. This is dictated by the intrinsically quantum nature of superfluidity.

To demonstrate this, let's consider a bucket filled with superfluid helium and examine the circulation κ around a closed contour L_1, as schematically shown in Fig. 1.4 (left). The circulation is defined as

$$\kappa = \oint_{L_1} \vec{v}_s \cdot d\vec{l}. \tag{1.4}$$

Here, $\vec{v}_s$ is the velocity of the superfluid flow. In Sect. 1.3 we have seen that $\vec{v}_s$ is proportional to the gradient of the phase of the macroscopic wave function $\psi(\vec{r}, t)$ describing BEC of helium atoms in the He II phase (see Eq. (1.1)). Plugging Eq. (1.2) into Eq. (1.4), we can write

$$\kappa = \frac{\hbar}{m_{\text{He}}} \oint_{L_1} \nabla S \cdot d\vec{l} = \frac{\hbar}{m_{\text{He}}} (\Delta S)_{L_1}, \tag{1.5}$$

where $(\Delta S)_{L_1}$ is the change in the phase after completing one loop around the contour L_1.

The macroscopic wave function $\psi(\vec{r}, t)$ is single-valued and, thus, upon the completion of a full trip around the closed contour L_1 in Fig. 1.4 (left), $\psi(\vec{r}, t)$ should stay unchanged. This results in a restriction on the wave function phase $S(\vec{r}, t)$, which is allowed to change only by an integral multiple of 2π. As can be seen from Fig. 1.4 (left), the fluid in the container occupies a simply-connected domain, i.e. the contour L_1 can be reduced to a single point. Then the phase $S(\vec{r}, t)$ can take any value in the range from 0 to 2π at the point of the L_1 reduction. But this would again violate the requirement for the wave function to be single-valued. The only solution to avoid this is to set the amplitude of the wave function to zero at the L_1 reduction point. Therefore, this implies the absence of the condensate at the point where $\psi(\vec{r}, t) = 0$ and, hence, the absence of the He II phase at that point. This naturally leads to the idea that the domain occupied by helium in its He II phase should be multiply-connected in order to ensure that the condensate wave function is single-valued. A multiply-connected domain is illustrated in Fig. 1.4 (right). Contour L_2 now encloses the obstacle ('hole') in the domain and describes a closed path fully passing through the superfluid . The superfluid circulation around contour L_2 is not equal to zero. Given that the wave function phase $S(\vec{r}, t)$ can take only integer multiples of 2π, the circulation given by Eq. (1.5) around contour L_2 enclosing the obstacle is quantized [3, 73]:

$$\kappa = n \frac{h}{m_{\text{He}}}, \tag{1.6}$$

where n is an integer 0, 1, 2, ...; h is the Planck constant. For $n = 1$, κ is known as the *quantum of circulation* and is equal to $\kappa = 9.98 \times 10^{-8}$ m^{2}s^{-1}. In superfluids, the obstacle in Fig. 1.4, from which the superfluid component is excluded, is called a vortex core and manifests as a region comprised of only the normal-fluid component. In superfluid ^{4}He the size of a vortex core is $\sim$1 Ångström [3].

It is interesting to note that there is a qualitative distinction between classical and quantum vortices imposed by the quantum mechanics: the circulation of vortices in superfluids is quantized, which makes them stable entities immune to a dissociation into any smaller constituents; whereas classical vortices are subject to inevitable

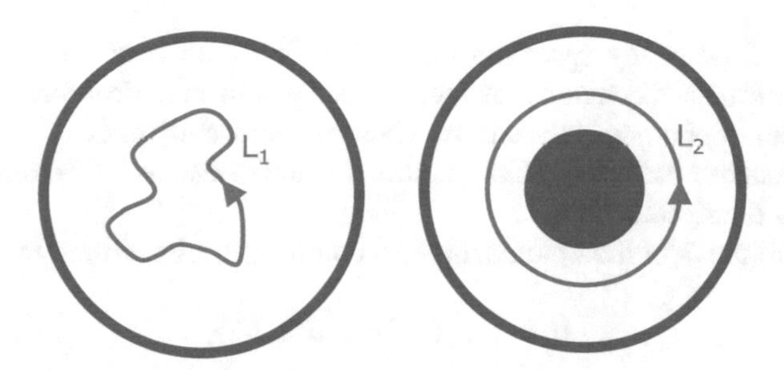

Fig. 1.4 Contour L_1 in a simply-connected domain (left) can be reduced to a single point. Contour L_2 enclosing an obstacle in a multiply-connected domain (right) cannot be reduced to a single point and leads to the quantization of the superfluid circulation

diffusive effects of viscosity, which makes them prone to a division into smaller vortices. Having said that, we would like to point out that quantum turbulence is a phenomenon immensely hard for full theoretical description and is a subject of an intensive ongoing research [74, 75].

1.8 A Glance at Optomechanics

The field of optomechanics explores the coupling and energy exchange between light and mechanical degrees of freedom [76–79]. While the mechanism of the interaction between optical field and a thin film of superfluid helium is elaborated in Sect. 2.2, here we introduce a generic optomechanical system and define its most important figures of merit. An archetypal set-up consists of a laser-driven Fabry-Perot optical cavity one mirror of which is mechanically compliant and the other one is fixed (Fig. 1.5). Fundamentally, the optomechanical coupling arises through the

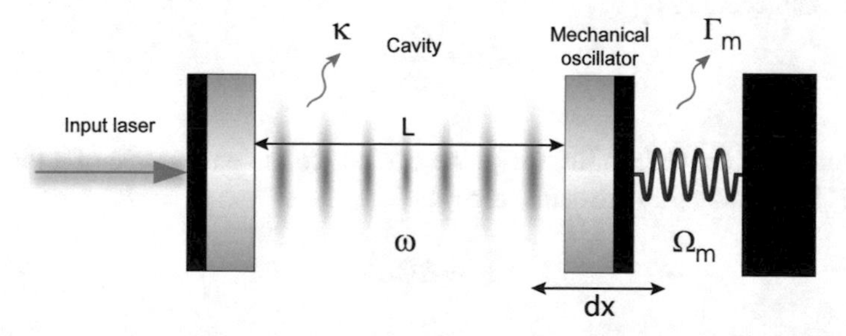

Fig. 1.5 An archetypal optomechanical system comprised of a Fabry-Perot optical cavity one mirror of which is mechanically compliant and the other one is fixed. Displacement of the moveable mirror is denoted as $\mathrm{d}x$. The optical and mechanical resonance frequencies are respectively given by ω and Ω_m and the corresponding dissipation rates as κ and Γ_m. L is the cavity length

modulation of the cavity spectrum induced by the motion of the compliant end-mirror. In return, the mechanical oscillator is subject to an optical force that depends on the number of photons in the cavity. Thus, the intracavity optical field both drives the mechanical oscillator via radiation pressure and measures its motion through the amplitude or phase detection.

The Hamiltonian of the optomechanical coupling can be written as [80]

$$\hat{H} = \hbar\omega(\hat{x})\hat{a}^\dagger\hat{a} + \hbar\Omega_\mathrm{m}\hat{b}^\dagger\hat{b}, \tag{1.7}$$

where ω and Ω_m are respectively the optical and mechanical frequencies; $\hat{a}$ ($\hat{b}$) and $\hat{a}^\dagger$ ($\hat{b}^\dagger$) are respectively the photon (phonon) annihilation and creation operators. Terms describing the laser driving and optical and mechanical dissipation are omitted. The optical resonance frequency is a function of the position of the compliant end-mirror which is given by $\hat{x} = x_\mathrm{zpf}\left(\hat{b} + \hat{b}^\dagger\right)$. Here, x_zpf is the root-mean-square displacement due to quantum-mechanical zero-point fluctuations which read [76]:

$$x_\mathrm{zpf} = \sqrt{\frac{\hbar}{2m_\mathrm{eff}\Omega_\mathrm{m}}}, \tag{1.8}$$

where m_eff is the effective mass of the oscillator. Typically, x_zpf is of the order of femtometers.

The resonance frequency of the cavity is given by $\omega = \frac{m\pi c}{L}$, where L is the cavity length and m is the optical mode number. In the limit when the mirror displacement dx is much smaller than L, the frequency ω can be Taylor-expanded to the first order in the displacement of the moveable mirror:

$$\omega\left(\hat{x}\right) \approx \omega_\mathrm{cav} - G\hat{x}, \tag{1.9}$$

where ω_cav is the cavity resonance frequency for $x = 0$. Here, we introduced the optomechanical coupling coefficient G, which shows the optical resonance frequency shift per displacement:

$$G = \left.\frac{\partial\omega}{\partial x}\right|_{x=0} \tag{1.10}$$

Plugging the above into Eq. (1.7), we can notice the emergence of a term $-\hat{F}\hat{x}$, where $\hat{F}$ is the radiation pressure force:

$$\hat{F} = \hbar G\hat{a}^\dagger\hat{a}. \tag{1.11}$$

Thus, we arrive at the standard optomechanical Hamiltonian, which reads [80]:

$$\hat{H} = \hbar\omega_\mathrm{cav}\hat{a}^\dagger\hat{a} + \hbar\Omega_\mathrm{m}\hat{b}^\dagger\hat{b} - \hbar g_0\hat{a}^\dagger\hat{a}\left(\hat{b} + \hat{b}^\dagger\right). \tag{1.12}$$

Here we defined the optomechanical single-photon coupling rate g_0 as

$$g_0 = G x_{\mathrm{zpf}}. \tag{1.13}$$

It shows the optical resonance frequency shift induced by a displacement equal to x_{zpf}.

Since most of the experiments can be described within the regime of linearized optomechanical interactions, we next perform the Hamiltonian linearization. To do this, we assume a bright pump laser field of the form $\hat{a} = (\alpha + \delta\hat{\alpha})e^{-i\omega_{\mathrm{L}}t}$, where $\alpha = \sqrt{N_{\mathrm{ph}}}$ is the amplitude of the intracavity optical field and $\delta\hat{\alpha}$ represents field fluctuations operator. Then we can write the photon number as

$$\hat{a}^{\dagger}\hat{a} = \alpha^2 + \alpha\left(\delta\hat{\alpha} + \delta\hat{\alpha}^{\dagger}\right) + \delta\hat{\alpha}^{\dagger}\delta\hat{\alpha}. \tag{1.14}$$

The linearization is performed by neglecting the product of the fluctuation operators—the last term (it is much smaller than the other terms). We also omit the first term as it represents just a classical static force. Going into a frame rotating at the laser frequency, we arrive at the Hamiltonian of linearized optomechanics [80]:

$$\hat{H} \approx \underbrace{-\hbar g_0\alpha\left(\delta\hat{\alpha}^{\dagger} + \delta\hat{\alpha}\right)\left(\hat{b} + \hat{b}^{\dagger}\right)}_{\text{linearized optomechanical interaction}} + \hbar\Omega_{\mathrm{m}}\hat{b}^{\dagger}\hat{b} - \hbar\Delta\delta\hat{\alpha}^{\dagger}\delta\hat{\alpha}, \tag{1.15}$$

where $\Delta \equiv \omega_{\mathrm{L}} - \omega_{\mathrm{cav}}$ is the laser detuning from the cavity resonance. We can also introduce the optomechanical coupling strength $g = g_0\alpha$, which scales with the number of intracavity photons and can be tuned by the amplitude of the pump laser. We also define one of the most important figures of merit of an optomechanical system—cooperativity, which is given by

$$C = \frac{4\,g^2}{\kappa\,\Gamma_{\mathrm{M}}}, \tag{1.16}$$

with κ being the optical decay rate. The cooperativity compares how fast the energy between optical and mechanical degrees of freedom is coherently exchanged relative to how fast the energy is lost via the optical and mechanical dissipation channels [76].

Optomechanical systems are known to exhibit an exquisite sensitivity to external perturbations which is broadly exploited for sensing of various fields and matter, from detection of gravitational waves [81] to weighing of a single molecule [82]. This thesis demonstrates how the exquisite sensitivity of an optomechanical system facilitates an investigation of the microscopic dynamics of strongly interacting quantum fluids, such as two-dimensional superfluid helium.

1.9 Whispering-Gallery-Mode Optical Microresonators

In order to boost the coherent energy exchange between optical and mechanical degrees of freedom and to enhance the cooperativity of the system (Eq. (1.16)), optical and mechanical dissipation rates should be minimized. This can be achieved by optimizing the quality of both optical and mechanical resonators. While Chap. 2 elaborates on the latter, here we discuss microcavities that provide optical modes with exceptionally low decay rates—whispering-gallery-mode resonators (WGM) [83, 84].

Optical whispering-gallery modes received their names in analogy with travelling acoustic waves observed in the whispering gallery of St. Paul's Cathedral in London and described by Lord Rayleigh [85]. WGM optical resonators are dielectric or semiconductor structures that confine light to small volumes for an extended amount of time [86]. Such dielectric structures generally have a circular or elliptical shape and guide light in closed paths via total internal reflection. Optical energy is typically confined to the periphery of the resonator. Resonances occur if an integer multiple of the light wavelength matches the effective path length along the circumference of the resonator. Then the photon lifetime in the cavity is enhanced through constructive interference. The resonance condition can be written as

$$2\pi R n_{\text{eff}} \simeq m\lambda, \tag{1.17}$$

where R is the resonator radius, n_{eff} its effective refractive index, and λ is the vacuum wavelength of light confined in the cavity. The integer m is the azimuthal order of the WGM and determines the number of nodes of its electric field along the resonator circumference. Solving Maxwell's equations in a circular domain with appropriate boundary conditions can yield degenerate WGM modes with the same azimuthal order m [87]. However, these modes differ in the number of their nodes along the resonator radius—radial WGM order is denoted by the integer p. The mode with $p = 1$ has its optical energy confined close to the resonator's outer boundary.

WGM optical resonators offer a number of desirable properties for optomechanical applications:

- Low optical decay rates and, hence, high quality factors [86]. In WGM resonators two major contributors to the optical dissipation—absorption and scattering—can be kept very low through respectively the selection of appropriate resonator materials and atomically-smooth interfaces. Optical quality factors of WGM resonators can be as high as 1.2×10^9 [88];
- High refractive index of the resonator material allows small mode volumes (typically sub-μm^3);
- Easy optical coupling through the evanescent field from an optical waveguide placed in the proximity to the resonator;
- WGM resonators simultaneously confine optical field and sustain a variety of in- and out-of-plane mechanical modes. The small optical mode volume allows a

large overlap between optical and mechanical energies, resulting in an optimized optomechanical coupling.

A combination of high quality factors and small mode volumes enhances light-matter interactions at interfaces of WGM resonators and makes them highly sensitive to any external perturbations occurring in the vicinity of the confined optical mode. For instance, WGM resonators are used for biosensing applications [89]. Moreover, the enhanced light-matter interactions and large optomechanical coupling rates make WGM resonators highly suitable candidates for investigating dynamics of quantum fluids in real time. This determines our choice of WGM resonators for the exploration of microscopic properties of thin films of superfluid helium presented in this thesis.

Various shapes and sizes of WGM resonators have been demonstrated over the past two decades. These include, but are not limited to, microtoroids [90, 91], silicon [92] and GaAs [93] microdisks, double-disk resonators [94], microspheres [95, 96], etc.

For our optomechanical configuration with thin films of superfluid helium, described in detail in Chap. 2, we typically utilize either silica microtoroids (Fig. 1.6(top left)) or microdisks (Fig. 1.6(top right)). These devices are fabricated from a thick silicon wafer bearing a thin thermal oxide layer (silica). Disks are defined in the silica layer by the means of photolithography and hydrofluoric acid (HF) wet-etch. These steps are followed by a XeF_2 gas-phase selective etching of the silicon material which leaves the silica disks isolated from the substrate and on top of a silicon pedestal (Fig. 1.6). Microtoroids are made from the disks through a CO_2 laser reflow which ensures atomically-smooth inner surface of the cavity with minimized optical losses and an enhanced quality factor. Fabricated devices show a number of WGM families, with optical quality factors in the range of 10^5–10^6 for microdisks and $\sim 10^7$ for microtoroids.

We would like to point out that silica is an excellent choice for the WGM resonators used for the experiments presented in this thesis, as the thermal oxide layer inherits the pristine smoothness of commercially grown single-crystal silicon wafers. In particular, the atomically-smooth upper surface of the resonators minimizes vortex pinning effects, allowing us to study coherent vortex dynamics in two-dimensional superfluid helium (see Sect. 5.11).

1.10 Third Sound

The mechanical degree of freedom in the optomechanical system presented in this thesis is provided by the motion of a thin film of superfluid helium. Such motion is called *third sound* (see Sect. 1.6) and constitutes fluctuations of the film thickness which propagate like a surface wave [21, 97–103]. In the case of thin films (typical thickness is $d \approx 1$ to 20 nm) the normal fluid component is clamped to the substrate due to its viscosity μ, as the viscous penetration depth at temperatures below the superfluid phase transition is $\delta = \sqrt{\frac{2\mu}{\rho\Omega_3}} \approx 200\ \mu\text{m}$ to 200 nm (for third-sound frequencies $\Omega_3/2\pi = 1$ Hz to 1 MHz), i.e. $\delta \gg d$. Therefore, only the superfluid

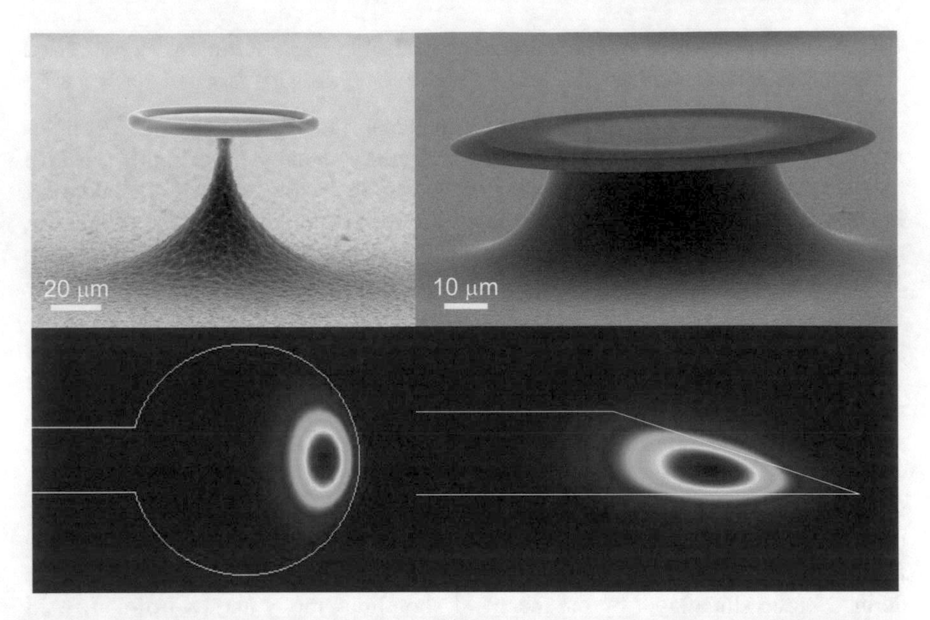

Fig. 1.6 Scanning-electron-microscope (SEM) images of a WGM microtoroid (top left) and a microdisk (top right), which are typically used in our experiments exploring the microscopic properties of thin films of superfluid helium. Intensity profiles of the lowest-order whispering-gallery modes supported by each of the resonators are shown under their SEM images

component sustains mechanical motion and, since in most experimental conditions the third-sound wavelength λ_3 is much greater than the film thickness d, oscillates parallel to the substrate. Third-sound waves have been observed in films as thin as only a few atomic layers [102].

A sketch of a third-sound wave is shown in Fig. 1.7a. Wave's crests contain a larger fraction of superfluid than equilibrium. And since the superfluid component does not carry entropy, i.e. no thermal energy (see Sect. 1.4), this results in a lowered temperature on the crest. In contrast, troughs contain a smaller fraction of superfluid than equilibrium and, therefore, have higher temperature [3, 67]. The natural flow of the superfluid component towards heated troughs via the fountain effect (see Sect. 1.5) drives the third-sound wave propagation. Moreover, since the troughs are hotter, helium atoms evaporate and recondense onto the colder crests. The third-sound wave is, therefore, naturally accompanied by temperature fluctuations [3]. Figure 1.7b illustrates the flow profile of the third-sound wave with directions of motion of the constituent helium atoms (red arrows). Since atoms under the crests move in-phase with the film surface deformation, their velocity is positive, whereas velocity of atoms under the troughs is negative. The condition of irrotationality of the superfluid flow in the third-sound wave, i.e. $\oint \vec{v} \cdot \mathrm{d}\vec{l} = 0$ for any closed loop inside the superfluid, demands that velocities of all helium atoms under the wave deformation profile are equal in magnitude, i.e. third-sound velocity $\vec{v}_3$ does not depend on z. Therefore, third-sound wave is associated with a net mass flow: at any

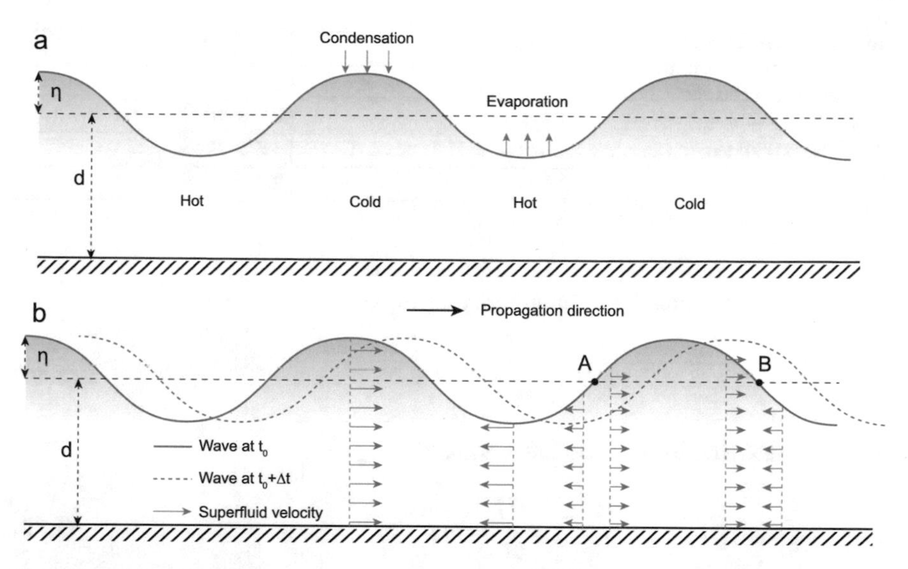

Fig. 1.7 **a** Sketch of a third-sound wave. d—mean superfluid film thickness, η—amplitude of the thickness deflection from equilibrium. Troughs contain a smaller fraction of the superfluid component than equilibrium and are, therefore, associated with raised temperature (because superfluid does not carry thermal energy). In contrast, crests contain more superfluid than equilibrium and are thereby colder. Helium atoms evaporate from the troughs and recondense onto the crests. **b** Flow profile of a third-sound wave. All atoms under the film surface have the same velocity magnitude. Atoms under the crests move in the direction of the wave propagation, whereas those under the troughs move in the opposite direction. Since there are more atoms under the crests than under the troughs, the third-sound wave is associated with a net mass flow

given time there is more fluid moving rightwards under the peaks than leftwards under the troughs (for the direction of the wave propagation in Fig. 1.7b). It is this net mass flow of the third-sound wave that couples to a vortex flow field, resulting in a vortex-induced third-sound mode frequency splitting (see Chap. 4).

Third-sound waves are identical to surface waves on a shallow water, although with a crucial difference: the restoring force for water waves is provided by gravity [104], whilst for third-sound it is provided by an attractive van der Waals interaction between helium atoms in the superfluid film and atoms of the substrate. The van der Waals force per unit mass is given by

$$F_{\text{vdw}} = \frac{3\,\alpha_{\text{vdw}}}{d^4}, \tag{1.18}$$

where α_{vdw} is the van der Waals coefficient characterizing the strength of the interaction between the film and the substrate. Van der Waals coefficients vary for different materials; as an example, a few of these coefficients are provided in Table 1.1.

Equations of motion governing the dynamics of third-sound waves are analogous to those for a classical liquid. For an inviscid and incompressible fluid, and in the limits of very long wavelength $d \ll \lambda_3$ and small wave amplitudes $\eta \ll d$, these

Table 1.1 Van der Waals coefficients for four sample materials which can serve as a substrate for a thin film of superfluid helium [105]

Material	α_{vdw}, $\mathrm{m^5 s^{-2}}$
Silica	2.6×10^{-24}
CaF_2	2.2×10^{-24}
Silicon	3.5×10^{-24}
MgO	2.8×10^{-24}

equations are presented by the continuity equation

$$\frac{\partial \eta}{\partial t} = -d\, \vec{\nabla} \cdot \vec{v}, \tag{1.19}$$

and by a linearized form of the Euler equation

$$\frac{\partial \vec{v}}{\partial t} = -g\, \vec{\nabla} \eta, \tag{1.20}$$

where g is the linearized acceleration due to the attractive van der Waals force to the substrate [3, 97]. The continuity equation Eq. (1.19) represents the conservation of mass and states that fluid influx in a particular region should result in film thickness increase in that region [3, 21, 22]. The linearized Euler equation Eq. (1.20) derives from Newton's second law and constitutes momentum conservation.

1.10.1 Third-Sound Mode Profiles

Assuming oscillatory solutions to Eqs. (1.19) and (1.20) and eliminating from them velocity $\vec{v}$, we can obtain a wave equation for the third sound:

$$\nabla^2 \eta + k^2 \eta = 0, \tag{1.21}$$

where k is the wave number. For resonators with cylindrical symmetry (such as WGM resonators), relevant to the experiments presented in this thesis, solutions to the wave equation $\eta(r, \theta)$ can be found via separation of variables in cylindrical coordinates (r, θ) [103, 106]. A general solution for the third-sound mode profile (r, θ) is dictated by the confining geometry. For circular resonators, it is based on the Bessel function of the first kind and is given as [21, 22, 103]

$$\eta\,(r, \theta, t) = \eta_0\, J_{\mathrm{m}} \left(\xi_{\mathrm{m,n}} \frac{r}{R} \right) e^{i(m\theta - \Omega_3 t)}. \tag{1.22}$$

Here, J_{m} is the Bessel function of the first kind of order m, $\xi_{\mathrm{m,n}}$ is a frequency parameter which depends on the mode order and boundary conditions; m and n are azimuthal and radial mode numbers respectively; η_0 is the mode amplitude; and R

is the resonator radius. Modes with $m = 0$ are rotationally invariant, whereas those with $m > 0$ are rotationally non-invariant. Typically, rotationally invariant modes exhibit the largest optomechanical coupling [93].

The third-sound phase velocity c_3 can also be identified from the wave equation Eq. (1.21) and is given by [97]

$$c_3 = \sqrt{\frac{\rho_s}{\rho} F_{\mathrm{vdw}}\, d} = \sqrt{3 \frac{\rho_s}{\rho} \frac{\alpha_{\mathrm{vdw}}}{d^3}}, \tag{1.23}$$

where ρ_s/ρ is the ratio of the superfluid to total fluid density. The velocity c_3, combined with appropriate boundary conditions, defines the third-sound frequency (given below as an angular mechanical frequency):

$$\Omega_3 = k_{\mathrm{m,n}}\, c_3 = \xi_{\mathrm{m,n}} \frac{c_3}{R}. \tag{1.24}$$

Here, $k_{\mathrm{m,n}} = \xi_{\mathrm{m,n}}/R$ is the wave number.

1.10.2 Boundary Conditions

The type of confinement provided by the resonator determines boundary conditions for the third-sound wave. These conditions are distinguished as "fixed"—*Dirichlet*—and "free"—*Neumann* (Fig. 1.8). The general solution to the wave equation for the third sound (Eq. (1.22)) is valid for both types of the boundary conditions, however yields distinctly different characteristic mode shapes for the two cases.

The fixed boundary condition implies zero superfluid displacement at the periphery of the resonator, i.e. $\eta(R, \theta, t) = 0$ (Fig. 1.8 left). This, in turn, requires the Bessel function to be $J_{\mathrm{m}}\left(\xi_{\mathrm{m,n}} \frac{r}{R}\right) = 0$ and, hence, the argument $\left(\xi_{\mathrm{m,n}} \frac{r}{R}\right)$ has to be a zero of J_{m}. Therefore, the frequency parameter $\xi_{\mathrm{m,n}}$ has to correspond to zeros of J_{m}, such that $\xi_{\mathrm{m,n}}$ is the nth zero (e.g. $\xi_{0,1} = 2.405$, $\xi_{1,3} = 10.174$, $\xi_{5,7} = 28.627$, etc. See Table 1.2). The fixed boundary condition allows fluid flow in and out of the resonator and, therefore, does not conserve volume. The superfluid film thickness at the boundary is fixed to the equilibrium value (Fig. 1.8 left).

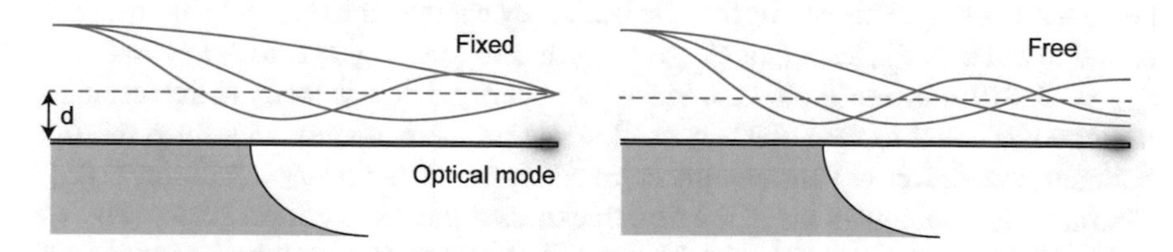

Fig. 1.8 Radial profiles $\eta(r)$ for the first three rotationally invariant ($m = 0$) third-sound modes with fixed (left) and free (right) boundary conditions. d is the mean superfluid film thickness. The glow represents an optical mode confined to the periphery of the resonator. The superfluid film and the WGM resonator are not drawn to scale

Table 1.2 First eight values of the frequency parameter ξ and frequencies for $(m = 1, n)$ third-sound modes. In the case of the fixed boundary conditions the frequency parameter ξ corresponds to zeros of the Bessel function J_{m}, whereas for the free boundary conditions ξ' corresponds to zeros of the Bessel function derivative J'_{m}. Third-sound frequencies $\Omega_3/2\pi$ are computed for a resonator with the following parameters: radius $R = 30$ µm, film thickness $d = 5$ nm

Fixed	$n = 1$	$n = 2$	$n = 3$	$n = 4$	$n = 5$	$n = 6$	$n = 7$	$n = 8$
$\xi_{1,\mathrm{n}}$	3.832	7.016	10.174	13.324	16.471	19.616	22.760	25.904
$\Omega_3/2\pi$, kHz	162.114	296.819	430.424	563.705	696.847	829.917	962.945	1095.95
Free	$n = 1$	$n = 2$	$n = 3$	$n = 4$	$n = 5$	$n = 6$	$n = 7$	$n = 8$
$\xi'_{1,\mathrm{n}}$	1.841	5.331	8.536	11.706	14.864	18.016	21.164	24.311
$\Omega_3/2\pi$, kHz	77.897	225.565	361.159	495.263	628.856	762.210	895.432	1028.58

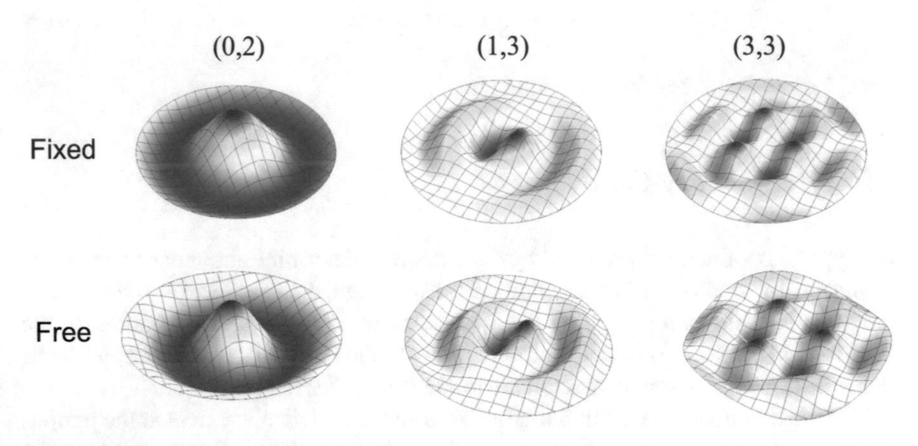

Fig. 1.9 Displacement profiles of a few sample third-sound modes on a circular resonator with fixed and free boundary conditions. Mode orders are shown above of each of the modes

The free boundary condition requires the radial derivative of the superfluid displacement, i.e. radial velocity of the superfluid flow, to be zero at $r = R$, i.e. $\partial_R \eta(R, \theta, t) = 0$ [103] (Fig. 1.8 right). Hence, the derivative of the Bessel function must also be zero, i.e. $J'_{\mathrm{m}}\left(\xi_{\mathrm{m,n}} \frac{r}{R}\right) = 0$, and that is ensured by the argument being a zero of J'_{m}. This entails that the frequency parameter in this case has to correspond to zeros of J'_{m}, such that $\xi'_{\mathrm{m,n}}$ is the nth zero (e.g. $\xi_{0,1} = 3.832$, $\xi_{1,3} = 8.536$, $\xi_{5,7} = 27.010$, etc. See Table 1.2). Since the superfluid flow velocity at the resonator periphery is equal to zero, there is no flow across the boundary and, thus, the free boundary condition is volume conserving ("no-flow" boundary condition [103]). This also allows fluctuations of the film thickness at the resonator periphery (Fig. 1.8 right). Displacement profiles of a few sample third-sound modes on a circular resonator with both fixed and free boundary conditions are shown in Fig. 1.9.

Boundary conditions for superfluid third-sound resonators are not typically known from first principles and usually determined from the measured mode spectrum frequencies (e.g. like in Sect. 5.4). Both fixed [15, 103, 106] and free [21–23, 107] boundary conditions have been reported in the literature.

References

1. Kapitza P (1938) Viscosity of liquid helium below the λ-point. Nature 141(3558):74
2. Allen JF, Misener AD (1938) Flow phenomena in liquid Helium II. Nature 142(3597):643–644
3. Tilley DR, Tilley J (1990) Superfluidity and superconductivity. CRC Press
4. Annett JF (2004) Superconductivity, superfluids, and condensates. Oxford University Press
5. Mermin ND, Wagner H (1966) Absence of ferromagnetism or antiferromagnetism in one- or two-dimensional isotropic Heisenberg models. Phys Rev Lett 17(22):1133–1136
6. Hohenberg PC (1967) Existence of long-range order in one and two dimensions. Phys Rev 158(2):383–386
7. Bishop DJ, Reppy JD (1978) Study of the superfluid transition in two-dimensional He 4 films. Phys Rev Lett 40(26):1727
8. Berezinskii VL (1971) Destruction of long-range order in one-dimensional and two-dimensional systems having a continuous symmetry group I. Classical systems. Sov J Exp Theor Phys 32(3):493–500
9. Berezinskii VL (1972) Destruction of long-range order in one-dimensional and two-dimensional systems possessing a continuous symmetry group. II. Quantum systems. Sov J Exp Theor Phys 34(3):610–616
10. Kosterlitz JM, Thouless DJ (1972) Long range order and metastability in two dimensional solids and superfluids. (Application of dislocation theory). J Phys C Solid State Phys 5(11):L124
11. Kosterlitz JM, Thouless DJ (1973) Ordering, metastability and phase transitions in two-dimensional systems. J Phys C Solid State Phys 6(7):1181
12. Nobel Media (2016) Topological phase transitions and topological phases of matter. Advanced information. NobelPrize.org
13. Nobel Media (2016) Strange phenomena in matter's fatlands. Popular science background. NobelPrize.org
14. Simula T, Davis MJ, Helmerson K (2014) Emergence of order from turbulence in an isolated planar superfluid. Phys Rev Lett 113(16):165302
15. Hoffmann JA, Penanen K, Davis JC, Packard RE (2004) Measurements of attenuation of third sound: evidence of trapped vorticity in thick films of superfluid ^{4}He. J Low Temp Phys 135(3–4):177–202
16. Onsager L (1949) Statistical hydrodynamics. Nuovo Cimento Suppl 6(S2):279–287
17. Whitmore SC, Zimmermann W Jr (1968) Observation of quantized circulation in superfluid helium. Phys Rev 166(1):181
18. Bewley GP, Lathrop DP, Sreenivasan KR (2006) Superfluid helium: visualization of quantized vortices. Nature 441(7093):588
19. Fonda E, Meichle DP, Ouellette NT, Hormoz S, Lathrop DP (2014) Direct observation of Kelvin waves excited by quantized vortex reconnection. Proc Natl Acad Sci 111:4707–4710
20. Adams PW, Glaberson WI (1987) Vortex dynamics in superfluid helium films. Phys Rev B 35(10):4633–4652
21. Ellis FM, Luo H (1989) Observation of the persistent-current splitting of a third-sound resonator. Phys Rev B 39(4):2703–2706
22. Ellis FM, Li L (1993) Quantum swirling of superfluid helium films. Phys Rev Lett 71(10):1577–1580
23. Sachkou YP, Baker CG, Harris GI, Stockdale OR, Forstner S, Reeves MT, He X, McAuslan DL, Bradley AS, Davis MJ, Bowen WP (2019) Coherent vortex dynamics in a strongly interacting superfluid on a silicon chip. Science 366(6472):1480–1485

24. Desbuquois R, Chomaz L, Yefsah T, Léonard J, Beugnon J, Weitenberg C, Dalibard J (2012) Superfluid behaviour of a two-dimensional Bose gas. Nat Phys 8(9):645–648
25. Zwierlein MW, Abo-Shaeer JR, Schirotzek A, Schunck CH, Ketterle W (2005) Vortices and superfluidity in a strongly interacting Fermi gas. Nature 435(7045):1047–1051
26. Lagoudakis KG, Wouters M, Richard M, Baas A, Carusotto I, André R, Dang LS, Deveaud-Plédran B (2008) Quantized vortices in an exciton–polariton condensate. Nat Phys 4(9):706–710
27. Amo A, Pigeon S, Sanvitto D, Sala VG, Hivet R, Carusotto I, Pisanello F, Leménager G, Houdré R, Giacobino E, Ciuti C, Bramati A (2011) Polariton superfluids reveal quantum hydrodynamic solitons. Science 332(6034):1167–1170
28. Dominici L, Carretero-González R, Gianfrate A, Cuevas-Maraver J, Rodrigues AS, Frantzeskakis DJ, Lerario G, Ballarini D, Giorgi MD, Gigli G, Kevrekidis PG, Sanvitto D (2018) Interactions and scattering of quantum vortices in a polariton fluid. Nat Commun 9(1):1467
29. Eckel S, Lee JG, Jendrzejewski F, Murray N, Clark CW, Lobb CJ, Phillips WD, Edwards M, Campbell GK (2014) Hysteresis in a quantized superfluid 'atomtronic' circuit. Nature 506(7487):200–203
30. Matthews MR, Anderson BP, Haljan PC, Hall DS, Wieman CE, Cornell EA (1999) Vortices in a Bose-Einstein condensate. Phys Rev Lett 83(13):2498–2501
31. Wilson KE, Newman ZL, Lowney JD, Anderson BP (2015) In situ imaging of vortices in Bose-Einstein condensates. Phys Rev A 91(2):023621
32. Gauthier G, Reeves MT, Yu X, Bradley AS, Baker MA, Bell TA, Rubinsztein-Dunlop H, Davis MJ, Neely TW (2019) Giant vortex clusters in a two-dimensional quantum fluid. Science 364(6447):1264–1267
33. Johnstone SP, Groszek AJ, Starkey PT, Billington CJ, Simula TP, Helmerson K (2019) Evolution of large-scale flow from turbulence in a two-dimensional superfluid. Science 364(6447):1267–1271
34. Bloch I, Dalibard J, Zwerger W (2008) Many-body physics with ultracold gases. Rev Mod Phys 80(3):885–964
35. Bloch I (2008) Quantum gases. Science 319(5867):1202–1203
36. Makotyn P, Klauss CE, Goldberger DL, Cornell EA, Jin DS (2014) Universal dynamics of a degenerate unitary Bose gas. Nat Phys 10(2):116–119
37. Husmann D, Uchino S, Krinner S, Lebrat M, Giamarchi T, Esslinger T, Brantut J-P (2015) Connecting strongly correlated superfluids by a quantum point contact. Science 350(6267):1498–1501
38. Valtolina G, Burchianti A, Amico A, Neri E, Xhani K, Seman JA, Trombettoni A, Smerzi A, Zaccanti M, Inguscio M, Roati G (2015) Josephson effect in fermionic superfluids across the BEC-BCS crossover. Science 350(6267):1505–1508
39. Behrle A, Harrison T, Kombe J, Gao K, Link M, Bernier J-S, Kollath C, Köhl M (2018) Higgs mode in a strongly interacting fermionic superfluid. Nat Phys 14(8):781
40. Kovtun PK, Son DT, Starinets AO (2005) Viscosity in strongly interacting quantum field theories from black hole physics. Phys Rev Lett 94(11):111601
41. Burgess C (2008) String cosmology: cosmic defects in the lab. Nat Phys 4(1):11–12
42. The STAR Collaboration (2017) Global Λ hyperon polarization in nuclear collisions. Nature 548(7665):62–65
43. Bertone G, Tait TMP (2018) A new era in the search for dark matter. Nature 562(7725):51
44. Page D, Prakash M, Lattimer JM, Steiner AW (2011) Rapid cooling of the neutron star in Cassiopeia A triggered by neutron superfluidity in dense matter. Phys Rev Lett 106(8):081101
45. Chamel N (2017) Superfluidity and superconductivity in neutron stars. J Astrophys Astron 38(3):43
46. Wlazłowski G, Sekizawa K, Magierski P, Bulgac A, Forbes MM (2016) Vortex pinning and dynamics in the neutron star crust. Phys Rev Lett 117(23):232701
47. Thouless DJ, Anglin JR (2007) Vortex mass in a superfluid at low frequencies. Phys Rev Lett 99(10):105301

48. Trabesinger A (2007) Superfluid dynamics: vortices on the scales. Nat Phys 3(9):591
49. Cox T, Stamp PCE (2013) Inertial and fluctuational effects on the motion of a Bose superfluid vortex. J Low Temp Phys 171(5):459–465
50. Simula T (2018) Vortex mass in a superfluid. Phys Rev A 97(2):023609
51. Sonin EB (1997) Magnus force in superfluids and superconductors. Phys Rev B 55(1):485–501
52. Thompson L, Stamp PCE (2013) Vortex dynamics: quantum versus classical regimes. J Low Temp Phys 171(5):526–538
53. Cox T, Stamp PCE (2017) Quantum vortex dynamics: results for a 2-D superfluid. arXiv: 1706.09919
54. Thompson L, Stamp PCE (2012) Quantum dynamics of a Bose superfluid vortex. Phys Rev Lett 108(18):184501
55. Barenghi CF, Donnelly RJ, Vinen WF (eds) (2001) Quantized vortex dynamics and superfluid turbulence. Springer, New York
56. Wilson C, Ellis FM (1995) Vortex creation and pinning in high amplitude third sound waves. J Low Temp Phys 101(3):507–512
57. Ellis FM, Wilson CL (1998) Excitation and relaxation of film flow induced by third sound. J Low Temp Phys 113(3–4):411–416
58. Leggett AJ (2006) Quantum liquids: Bose condensation and Cooper pairing in condensed-matter systems. Oxford University Press
59. Leggett AJ (2008) Quantum liquids. Science 319(5867):1203–1205
60. Putterman SJ, Rudnick I (2008) Quantum nature of superfluid helium. Phys Today 24(8):39
61. Donnelly RJ (2009) The two-fluid theory and second sound in liquid helium. Phys Today 62(10):34–39
62. Tisza L (1938) Transport phenomena in Helium II. Nature 141(3577):913
63. London F (1938) The λ-phenomenon of liquid helium and the Bose-Einstein degeneracy. Nature 141(3571):643–644
64. Landau L (1941) The theory of superfluidity of Helium II. J Phys USSR 5(1–6):71–90
65. Landau LD, Lifshitz EM (2013) Fluid mechanics. Elsevier
66. Donnelly RJ, Barenghi CF (1998) The observed properties of liquid helium at the saturated vapor pressure. J Phys Chem Ref Data 27(6):1217–1274
67. Enss C, Hunklinger S (2005) Low-temperature physics. Springer
68. Allen JF, Jones H (1938) New phenomena connected with heat flow in Helium II. Nature 141:243–244
69. Kagan MY (2013) Modern trends in superconductivity and superfluidity. Lecture notes in physics. Springer, Netherlands
70. Peshkov V (1946) Determination of the velocity of propagation of the second sound in Helium II. J Phys USSR 10:389
71. Lane CT, Fairbank HA, Fairbank WM (1947) Second sound in liquid Helium II. Phys Rev 71(9):600–605
72. Saffman PG (1992) Vortex dynamics. Cambridge University Press
73. Glaberson WI, Schwarz KW (2008) Quantized vortices in superfluid Helium-4. Phys Today 40(2):54
74. Anderson BP (2016) Fluid dynamics: turbulence in a quantum gas. Nature 539(7627):36–37
75. Reeves M (2017) Quantum analogues of two-dimensional classical turbulence. PhD thesis, University of Otago
76. Aspelmeyer M, Kippenberg TJ, Marquardt F (2014) Cavity optomechanics. Rev Mod Phys 86(4):1391–1452
77. Aspelmeyer M, Meystre P, Schwab K (2012) Quantum optomechanics. Phys Today 65(7):29–35
78. Favero I, Marquardt F (2014) Focus on optomechanics. New J Phys 16(8):085006
79. Kippenberg TJ, Vahala KJ (2008) Cavity optomechanics: back-action at the mesoscale. Science 321(5893):1172–1176
80. Marquardt F (2011) Quantum optomechanics

81. LIGO Scientific Collaboration and Virgo Collaboration (2016) Observation of gravitational waves from a binary black hole merger. Phys Rev Lett 116(6):061102
82. Burg TP, Godin M, Knudsen SM, Shen W, Carlson G, Foster JS, Babcock K, Manalis SR (2007) Weighing of biomolecules, single cells and single nanoparticles in fluid. Nature 446(7139):1066–1069
83. Matsko AB, Ilchenko VS (2006) Optical resonators with whispering-gallery modes-part I: basics. IEEE J Sel Top Quantum Electron 12(1):3–14
84. Strekalov DV, Marquardt C, Matsko AB, Schwefel HGL, Leuchs G (2016) Nonlinear and quantum optics with whispering gallery resonators. J Opt 18(12):123002
85. Rayleigh OM (1910) The problem of the whispering gallery. Lond Edinb Dublin Phil Mag J Sci 20(120):1001–1004
86. Vahala K (2004) Optical microcavities. World Scientific
87. Baker C (2013) On-chip nano-optomechanical whispering gallery mode resonators. PhD Thesis, University Paris Diderot
88. Shitikov AE, Bilenko IA, Kondratiev NM, Lobanov VE, Markosyan A, Gorodetsky ML (2018) Billion Q-factor in silicon WGM resonators. Optica 5(12):1525–1528
89. Knittel J, Swaim JD, McAuslan DL, Brawley GA, Bowen WP (2013) Back-scatter based whispering gallery mode sensing. Sci Rep 3:2974
90. Armani DK, Kippenberg TJ, Spillane SM, Vahala KJ (2003) Ultra-high-Q toroid microcavity on a chip. Nature 421(6926):925
91. Kippenberg TJ, Spillane SM, Vahala KJ (2004) Demonstration of ultra-high-Q small mode volume toroid microcavities on a chip. Appl Phys Lett 85(25):6113–6115
92. Sun X, Zhang X, Tang HX (2012) High-Q silicon optomechanical microdisk resonators at gigahertz frequencies. Appl Phys Lett 100(17):173116
93. Ding L, Baker C, Senellart P, Lemaitre A, Ducci S, Leo G, Favero I (2010) High frequency GaAs nano-optomechanical disk resonator. Phys Rev Lett 105(26):263903
94. Wiederhecker GS, Chen L, Gondarenko A, Lipson M (2009) Controlling photonic structures using optical forces. Nature 462(7273):633–636
95. Gorodetsky ML, Savchenkov AA, Ilchenko VS (1996) Ultimate Q of optical microsphere resonators. Opt Lett 21(7):453–455
96. Carmon T, Vahala KJ (2007) Modal spectroscopy of optoexcited vibrations of a micron-scale on-chip resonator at greater than 1 GHz frequency. Phys Rev Lett 98(12):123901
97. Atkins KR (1959) Third and fourth sound in liquid Helium II. Phys Rev 113(4):962–965
98. Everitt CWF, Atkins KR, Denenstein A (1962) Detection of third sound in liquid helium films. Phys Rev Lett 8(4):161–163
99. Pickar KA, Atkins KR (1969) Critical velocity of a superflowing liquid-helium film using third sound. Phys Rev 178(1):389–399
100. Bergman D (1969) Hydrodynamics and third sound in thin He II films. Phys Rev 188(1):370–384
101. Bergman DJ (1971) Third sound in superfluid helium films of arbitrary thickness. Phys Rev A 3(6):2058–2066
102. Scholtz JH, McLean EO, Rudnick I (1974) Third sound and the healing length of He II in films as thin as 2.1 atomic layers. Phys Rev Lett 32(4):147–151
103. Schechter AMR, Simmonds RW, Packard RE, Davis JC (1998) Observation of 'third sound' in superfluid He-3. Nature 396(6711):554–557
104. Phillips OM (1969) The dynamics of the upper ocean. Cambridge University Press, London
105. Baker CG, Harris GI, McAuslan DL, Sachkou Y, He X, Bowen WP (2016) Theoretical framework for thin film superfluid optomechanics: towards the quantum regime. New J Phys 18(12):123025
106. Schechter AMR, Simmonds RW, Davis JC (1998) Capacitive generation and detection of third sound resonances in saturated superfluid He-4 films. J Low Temp Phys 110(1–2):603–608
107. Vorontsov A, Sauls JA (2004) Spectrum of third sound cavity modes on superfluid He-3 films. J Low Temp Phys 134(3–4):1001–1008

Chapter 2
Optomechanical Platform for Probing Two-Dimensional Quantum Fluids

In this chapter we present a paradigm of cavity optomechanics with thin films of superfluid helium. Two-dimensional superfluid films offer a number of desirable properties for both optomechanical protocols and investigation of quantum fluids. These properties include, but are not limited to, ultra-low mechanical dissipation and optical absorption, self-assembling nature of films on an optical cavity, strong coupling of superfluid mechanical excitations to quantized vortices, etc. We describe the mechanism of dispersive optomechanical coupling between an optical field confined within a whispering-gallery-mode resonator and mechanical vibrations of the superfluid film uniformly coating the resonator. The film naturally forms on surfaces due to a combination of ultra-low viscosity and attractive van der Waals forces. This is followed by a description of the experimental implementation of the optomechanical scheme with thin superfluid films. Enhanced light-matter interaction between confined optical field and superfluid film allowed us to resolve and track thermomechanical motion of superfluid helium in real time, which is the first demonstration of such capability to our best knowledge. This is crucial for understanding of both micro- and- macroscopic properties of superfluid helium. Moreover, utilizing the effect of dynamical backaction we also demonstrated the ability to laser cool and heat mechanical modes of the superfluid film.

2.1 Introduction

The field of cavity optomechanics studies the interaction between confined optical field and mechanical degrees of freedom [1, 2]. To date, the majority of optomechanical configurations have relied on mechanical motion of solid-state oscillators, such as thin membranes [3–5], micromirrors [6], nanorods [7], whispering-gallery-mode (WGM) resonators [8–10], photonic crystals [11], etc. Optomechanical methods applied to these structures enabled an exquisite control over mechanical degrees of freedom even at the quantum level, which led to such breakthroughs as cooling of

Y. Sachkou, *Probing Two-Dimensional Quantum Fluids with Cavity Optomechanics*, Springer Theses, https://doi.org/10.1007/978-3-030-52766-2_2

mechanical oscillators to their quantum ground states [12, 13], quantum squeezing of motion in mechanical resonators [14, 15], preparation and control of non-classical states of motion [16], entanglement between macroscopic mechanical oscillators [17, 18], etc.

Micro- and nanoscale mechanical oscillators actuated via their interaction with light enable applications of ultraprecise sensing of various physical quantities and systems [19–22]. In light of this, the utility of optomechanical applications has been extended to biological systems; for instance, for their detection and interrogation. The necessity to study biological objects, such as living cells, in their natural aqueous environments [23] has sparked a great interest in investigating optomechanical systems that interact with or are immersed in liquids [24–28]. For example, immersed optomechanics have made it possible to weigh single bacterial cells, nanoparticles, and proteins with sub-femtogram resolution [29]. However, viscous damping and substantially degraded mechanical quality factors of resonators in aqueous environments impose a constraint on the sensitivity and resolution of liquid-submerged optomechanical oscillators. In contrast, viscous damping can be dramatically reduced by confining a liquid inside the resonator, as opposed to immersing the resonator in liquid [30]. Optomechanical oscillations excited within hollow microfluidic resonators [31, 32] are sensitive to the change of properties of the liquid inside the cavity and have enabled, for instance, optical sensing of liquid density [33] and high-accuracy measurements of fluid viscosity [34].

In a reverse approach, liquids are used as resonators themselves. Surface tension shapes liquid droplets into impeccable spheres with atomically-smooth surface that can act as high-quality WGM resonators and co-host both optical and acoustic resonances [35]. For instance, droplets of water sustain both capillary oscillations [36, 37] and acoustic modes [38] which can be read out optically. An efficient interaction between acoustical and optical resonances within the droplet can result in stimulated Brillouin scattering, allowing the droplet to perform as a hypersound laser [39]. Moreover, droplets of other liquids, such as glycerol, have been used to perform high-resolution humidity sensing [40].

Typically, in order to maximize the rate of coherent exchange of energy between optical and mechanical degrees of freedom, optical and mechanical dissipation rates should be minimized. Then a naturally arising question is: what if a classical fluid that has viscosity and significant optical absorption is replaced with an inviscid quantum fluid that also has ultra-low optical dissipation, i.e. superfluid? Indeed, mechanical motion of superfluids has recently been identified as an attractive mechanical degree of freedom for optomechanical systems [41]. Superfluid helium-4 has been chosen as a suitable candidate for the role, and here is what makes it attractive:

- zero viscosity [42] results in ultra-low mechanical dissipation;
- $\sim$16 eV bandgap [43], i.e. it is transparent to any light wavelength above $\sim$80 nm—this determines its ultra-low optical absorption for the wavelength range of interest (either visible or infrared);
- has almost no impurities and no structural defects; and
- very high thermal conductivity.

The first demonstrated superfluid optomechanical system was designed by the group led by K. Schwab and was comprised of a superconducting niobium cell filled with superfluid helium [41]. Pressure waves in bulk helium ("first sound"—see Sect. 1.6) formed a gram-scale oscillator that was parametrically coupled to the niobium microwave resonator. Quality factors of superfluid mechanical resonances in the system reached values as high as 1.4×10^8 [44, 45]. Upon certain changes in the experimental configuration, this system is a promising platform for the detection of gravitational waves [46]. It has further been shown by the group of J. Davis that resonance frequencies of superconducting three-dimensional cavities can be tuned in situ at the percent level by varying the amount of superfluid helium in the cavity at millikelvin temperatures [47]. This can be useful for applications in superconducting quantum circuits and quantum cavity electromechanics.

The group led by J. Davis also demonstrated another example of a superfluid optomechanical system which boasts mechanical resonances with very high quality factors, namely the superfluid Helmholtz resonator with a capacitive detection scheme [48]. Upon a proper choice of material which the resonator is fabricated from, superfluid Helmholtz modes achieve ultra-low mechanical dissipation, reaching quality factors as high as $Q \sim 10^6$ [49]. A major advantage of the Helmholtz resonator over the superfluid optomechanical configuration demonstrated by the group of K. Schwab [41, 44] is that the superfluid Helmholtz modes exhibit effective masses 10^6 times smaller than superfluid modes in Refs. [41, 44], which makes the Helmholtz scheme more suitable for operating in the quantum regime of optomechanical interactions. The Helmholtz structure enabled measurements of superfluid density, mass flow and dissipation, and can also potentially serve for the detection of gravitational waves [49].

Another example of a superfluid optomechanical configuration was demonstrated by the group of J. Harris. The system consists of a miniature superfluid-filled optical cavity formed in between of a pair of single-mode optical fibres [50–52]. Bulk superfluid mechanical modes are electrostrictively coupled to the optical field confined inside the cavity. Near-perfect overlap between optical and mechanical modes results in a single-photon optomechanical coupling rate as high as 3 kHz. This makes this system promising for applications in quantum optomechanical protocols. Moreover, inspired by the implementation of classical-liquid droplets as optomechanical resonators [35], the group of J. Harris designed an optomechanical system based on a drop of superfluid helium suspended in vacuum by magnetic field. Theoretical calculations have already suggested that the efficient optomechanical coupling between optical and acoustic modes co-located within the droplet should enable a single-photon strong coupling regime [53]. The system has already been realized experimentally and is currently being investigated.

Two more groups active in the field of superfluid optomechanics are the group of H. Tang, which exploits optomechanical microdisk resonators submerged in bulk superfluid helium to investigate coupling between the resonator and phonons in the quantum fluid [54]; and the group of D. Zmeev, which utilizes doubly-clamped nanobeams as probes in superfluid helium [55, 56].

All incarnations of superfluid optomechanics described above utilize bulk helium. In contrast, we propose and present in this thesis an alternative approach based on two-dimensional ultra-light (pg–fg mass) films of superfluid helium-4 [57–60]. Thin superfluid films offer a range of desirable properties which make them a promising platform for cavity optomechanics:

- a combination of near-zero viscosity and attractive van der Waals forces [42] enable a self-assembling nature of the films on any surface, i.e. they in situ align themselves with an optical resonator of any shape;
- films exhibit very low mass (pg–fg range), minute volume and can be just a few atomic layers thick [61];
- film thickness and frequency of third-sound mechanical resonances on the film surface are largely tunable just by varying the pressure of helium gas injected in the sample chamber (see Sect. 2.3.2);
- third-sound modes exhibit low effective mass which enables large zero-point motion, single-photon optomechanical coupling and cooperativity (see Sect. 2.2.2) —figures of merit particularly sought after in the field of cavity optomechanics [1];
- strong phonon-phonon interactions result in nonlinearities (e.g. Duffing) of third-sound resonances which may lead to a generation of non-classical mechanical states;
- third-sound waves exhibit strong coupling to quantized vortices in the film, which enables real-time resolution of vortex dynamics (see Chap. 5);
- third-sound modes interact with electrons floating on the film surface, enabling a configuration for promising applications in circuit quantum electrodynamics and quantum computing [62, 63]; and
- thin superfluid films offer a prospect for applications such as superfluid matter-wave interferometers, gyroscopes, superfluid phononic circuits, etc.

In this chapter we first describe the mechanism of dispersive optomechanical coupling between an optical field confined within a WGM resonator and mechanical vibrations of the superfluid film uniformly coating the resonator. We show that the effective mass of mechanical modes of superfluid film, in contrast to intuition and effective mass of vibrational modes of solid-state membrane oscillators, scales as R^4/d, where R is the superfluid resonator radius and d is the film thickness. Thus, upon proper optimization of the resonator dimensions, film thickness and superfluid mechanical mode order, it is possible to achieve large amplitude of zero-point fluctuations and, hence, single-photon optomechanical coupling rate ($\frac{g_0}{2\pi} \gtrsim 100$ kHz [59]) and cooperativity ($C_0 \gtrsim 10$ [59]), which are desirable for the applications of quantum optomechanics [1].

Enhanced light-matter interaction at the interface of the high-quality optical WGM enabled by a small mode volume and strong evanescent optomechanical coupling enables ultra-high sensitivity of our optomechanical configuration to any perturbations occurring at the resonator surface, such as motion of the superfluid film. This opens up a great prospect for studying micro- and macroscopic properties of quantum fluids in regimes hitherto inaccessible. Thus, as the first demonstration of the utility of our scheme we study thermally driven motion of the superfluid helium film. Thermal elementary excitations—such as phonons and rotons—and interactions between them determine dynamics of superfluids, as well as their micro- and macroscopic properties, such as phase transitions [64], dissipation [65, 66], and quantum turbulence [67]. In order to understand these processes, it is crucial to be able to resolve and track superfluid thermal excitations in real time. Previous techniques to probe these fluctuations, such as neutron and light scattering [42, 68, 69], are slow compared to characteristic dissipation rates of the excitations, restricting them to measuring only average thermodynamic properties of superfluids. However, the sensitivity of our optomechanical technique allows us to resolve and track superfluid phonon excitations in real time. Moreover, utilizing the effect of dynamical backaction from both radiation pressure and photothermal forces, we demonstrate laser cooling and heating of mechanical modes of the superfluid film.

2.2 Optomechanical Interaction Between Optical Field and Superfluid Film

2.2.1 Interaction Mechanism

We begin with a description of the optomechanical[1] interaction between optical field and superfluid film. The coupling scheme is presented in Fig. 2.1. Light is confined to the periphery of a whispering-gallery-mode (WGM) optical cavity with cylindrical symmetry[2] coated with a few-nanometer thick film of superfluid helium [57] (Fig. 2.1a). As discussed in Sect. 1.10, mechanical waves existing in the superfluid film are known as third sound [70–72]. They manifest as ripples propagating on the surface of the film, as can be seen in Fig. 2.1a. At the periphery of the microcavity these ripples (i.e. film thickness fluctuations) perturb the evanescent field of the confined WGM and, thus, dispersively couple to it by modulating the effective refractive index of the material within the WGM's near field. This results in an alteration of the effective optical path length of the resonator and, hence, according to Eq. (1.17), leads to a shift of the WGM resonance frequency [59]. By utilizing a perturbation theory approach and comparing the total electromagnetic energy of the WGM to that fraction of it which is contained within the superfluid film, it is possible to estimate the frequency shift $\Delta\omega$ experienced by a WGM of resonance frequency ω_0 induced by the superfluid film [73, 74]:

$$\frac{\Delta\omega}{\omega_0} = -\frac{1}{2}\frac{\int_{\text{film}} (\varepsilon_{\text{sf}} - 1) \left|\vec{E}(\vec{r})\right|^2 \text{d}^3\vec{r}}{\int_{\text{all}} \varepsilon_r(\vec{r}) \left|\vec{E}(\vec{r})\right|^2 \text{d}^3\vec{r}}. \tag{2.1}$$

[1] This section incorporates a fraction of the work published by IOP Publishing: C. G. Baker, G. I. Harris, D. L. McAuslan, **Y. Sachkou**, X. He, W. P. Bowen. Theoretical framework for thin film superfluid optomechanics: towards the quantum regime. *New Journal of Physics* 18, 123025, 2016.

[2] Without loss of generality of the interaction mechanism here we consider a WGM microdisk; although the same coupling scheme applies to microtoroids, spheres, annular microdisks and other WGM optical cavities.

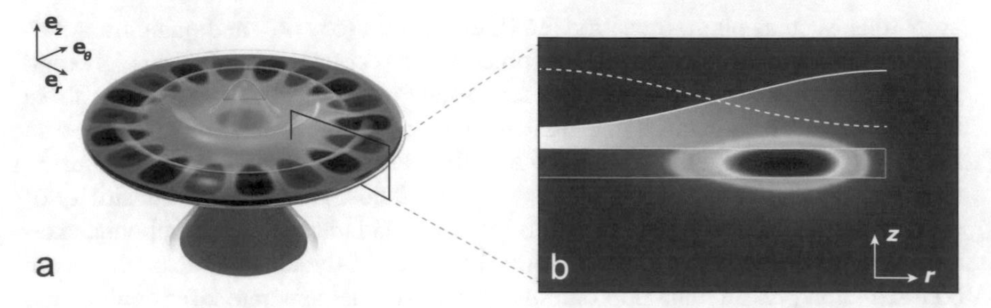

Fig. 2.1 **a** Artist's rendering of an optical whispering-gallery-mode microdisk coated with a few-nanometer thick film of superfluid helium. Red and blue lobes represent nodes and anti-nodes of the WGM optical field confined to the periphery of the microdisk. **b** Radial cross-section of the microdisk with the WGM optical field intensity simulated via finite-element method (FEM). Thickness fluctuations of the superfluid film driven by the third-sound wave (solid and dashed white lines) dispersively couple to the confined optical field inducing the WGM resonance frequency shift given by Eq. (2.1). Figure is reproduced from Ref. [59]

Here, $\varepsilon_r(\vec{r})$ is the relative permittivity of the material 'felt' by the unperturbed WGM electric field $\vec{E}(\vec{r})$, i.e. in the absence of the superfluid film. The relative permittivity of superfluid helium is given by $\varepsilon_{\mathrm{sf}} = 1.058$ [75]. The integral in the numerator is computed over the volume of the film, while the denominator integral is evaluated over the volume of all space. Thus, the electromagnetic energy of the WGM denoted in the numerator of Eq. (2.1) effectively "senses" the perturbing element (superfluid film) and its motion. Therefore, in order to optimize the light-superfluid optomechanical coupling, is it essential to maximize the evanescent WGM electric field at the resonator interface.

Figure 2.2a shows $\left|\vec{E}\right|^2$ of a transverse electric (TE) [76] WGM simulated via finite-element modelling (FEM). In this example the optical field is confined in a silica disk with thickness of 2 μm [77]. The solid black line indicates the spatial distribution of $\left|\vec{E}\right|^2$ along z-axis obtained at the disk cut passing through the WGM center (marked by the dashed line). As can be seen from this distribution, the WGM electric field is predominantly confined within the silica disk, whereas its evanescent tails on the top and bottom surfaces of the cavity, where the field couples to the superfluid film (red dashed lines), exhibit very low intensity. This results in weak WGM sensitivity/optomechanical coupling to the superfluid through Eq. (2.1). This makes the disk thickness one of the most crucial parameters for optimizing the WGM–superfluid interaction. Reduced thickness increases the WGM field evanescent tails outside of the resonator material, thus optimizing the overlap between $\left|\vec{E}\right|^2$ and the superfluid film. This leads to a much larger WGM resonance frequency shift caused by the film thickness fluctuations (see Eq. (2.1)). Figure 2.2b shows how the effective refractive index felt by the fundamental WGM mode varies with the disk thickness. If the disk is too thick and the optical mode resides predominantly inside the resonator, the effective refractive index is equal to that of silica. In contrast, if the disk is too thin the optical field becomes too delocalized along the z-axis, with

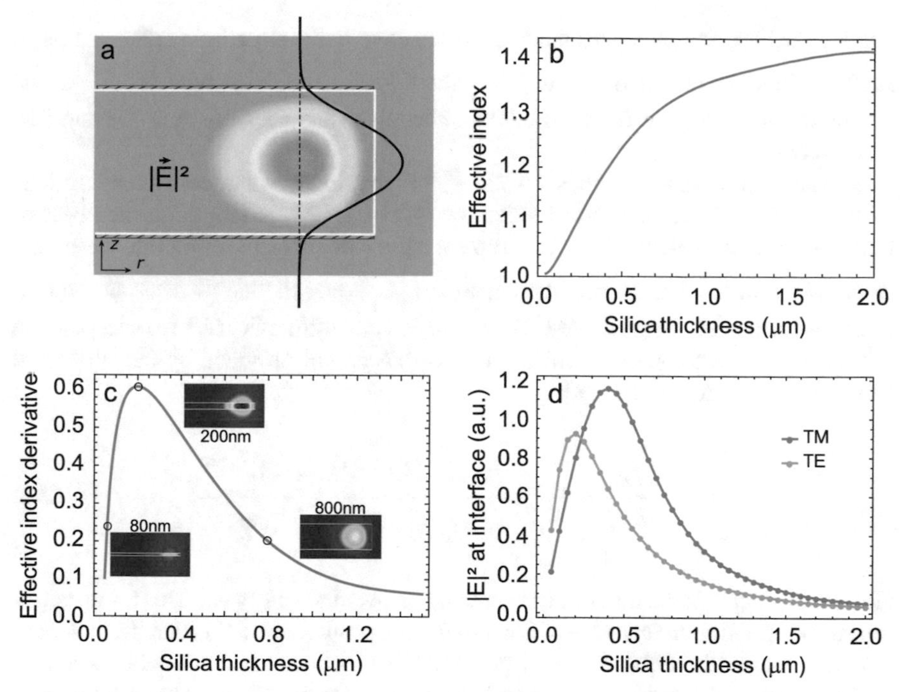

Fig. 2.2 **a** $\left|\vec{E}\right|^2$ of a transverse electric (TE) WGM simulated via finite-element modelling (FEM). The optical field is confined in a silica disk with thickness of 2 μm. The solid black line indicates the spatial distribution of $\left|\vec{E}\right|^2$ along z-axis obtained at the disk cut passing through the WGM center. The superfluid film is marked by the red dashed lines. **b** Effective refractive index of the silica microdisk as a function of the disk thickness. When the optical mode is completely contained inside the resonator, the effective refractive index will equal to that of silica. In contrast, when the optical mode is delocalized outside of the disk, the effective refractive index will equal to that of vacuum. **c** Derivative of the effective refractive index $\frac{dn_{eff}}{dt}$ as a function of the silica disk thickness. The maximum of $\frac{dn_{eff}}{dt}$ occurs at a thickness of 200 nm (for TE WGM). **d** Magnitude of $\left|\vec{E}\right|^2$ at the resonator interface as a function of the silica disk thickness for both TE and TM modes. Simulations performed by C. G. Baker

its value at the resonator interface substantially reduced, and the effective refractive index becomes equal to that of vacuum. The optical mode confined within the disk is most sensitive to an external dielectric placed at the surface of the resonator when the change in the effective refractive index with thickness is maximal. Figure 2.2c shows the derivative of the effective refractive index $\frac{dn_{eff}}{dt}$ as a function of the silica disk thickness. As can be seen, there is an optimal thickness when the WGM field is the most sensitive to perturbations caused by the superfluid film on the surface of the resonator. For a silica disk this occurs at 200 nm (for TE WGM), as seen in Fig. 2.2c. We would like to note that the optimal thickness also depends on the WGM polarization. In contrast to TE modes, which provide a continuous electric field across the disk interface, transverse magnetic (TM) modes exhibit a step increase in the field outside the disk, which makes them more sensitive to the superfluid at the resonator surface. This is illustrated in Fig. 2.2d, which shows the magnitude of $\left|\vec{E}\right|^2$ at the resonator interface as a function of the silica disk thickness for both TE and TM polarizations.

At the optimal disk thickness, when $\frac{\mathrm{d}n_{\mathrm{eff}}}{\mathrm{d}t}$ is maximal, evanescent decaying tails of the WGM field extend out of silica over a characteristic length on the order of hundreds of nanometers. Given that typical values of superfluid film thickness vary in the range $\Delta z = 1$ to 30 nm, the change in $\left|\vec{E}\right|^2$ over the length scale of the film thickness can safely be neglected. Therefore, we can utilize Eq. (2.1) to compute the WGM resonance frequency shift per unit displacement provided by the superfluid film, $G = \Delta\omega_0/\Delta z$ [1, 10, 78]:

$$G = \frac{\Delta\omega_0}{\Delta z} = -\frac{\omega_0}{2}\,\frac{\int_{\text{interface}} (\varepsilon_{\text{sf}} - 1)\left|\vec{E}(\vec{r})\right|^2 \mathrm{d}^2\vec{r}}{\int_{\text{all}} \varepsilon_r(\vec{r})\left|\vec{E}(\vec{r})\right|^2 \mathrm{d}^3\vec{r}}. \tag{2.2}$$

Here, the integral in the numerator is now taken over the cavity surface (for instance, top surface for simplicity). One of the remarkable properties of G is that it is independent of the superfluid film thickness and of the WGM cavity radius (in the same way as, for instance, for double-disk optomechanical resonators [79]). Through the optimization of the resonator thickness and WGM polarization, and taking into account a variation in the mean film thickness over the entire surface of the cavity (top, bottom, side), it is possible to achieve values of G as high as ~10 GHz/nm. This entails that variation in film thickness of only 10 pm (i.e. about $\frac{1}{36}$th of a helium monolayer [80]) would be enough to shift a WGM with optical quality factor of $Q = 2 \times 10^6$ by one linewidth. This equips us with an ultraprecise tool to optically characterize thickness of the superfluid film and dynamics of the mechanical excitations on its surface, which constitutes a substantial improvement comparing to capacitive detection schemes commonly exploited in the research with superfluids [72, 81].

2.2.2 *Optomechanical Coupling and Third-Sound Effective Mass*

In Sect. 2.2.1 we calculated the optomechanical coupling G which tells us how much the WGM optical resonance shifts per unit of the superfluid film thickness. We can now make a step further and calculate the single-photon optomechanical coupling g_0 and consequently the single-photon cooperativity C_0, two crucial figures of merit for quantum optomechanical systems [1] (see Sect. 1.8). The single-photon optomechanical coupling is given by

$$g_0 = G\,x_{\text{zpf}} = G\sqrt{\frac{\hbar}{2\,m_{\text{eff}}\,\Omega_{\text{M}}}}, \tag{2.3}$$

where x_{zpf} is the root-mean-square displacement due to quantum-mechanical zero-point fluctuations and Ω_{M} is the mechanical frequency of the oscillator. The single-photon cooperativity can then be calculated as

$$C_0 = \frac{4\, g_0^2}{\kappa\, \Gamma_{\mathrm{M}}}, \tag{2.4}$$

where κ is the optical decay rate (linewidth) of a WGM and Γ_{M} is the dissipation rate of the mechanical oscillator (third-sound modes in the case of superfluid films—see Sect. 1.10). As can be seen, computation of g_0 requires the knowledge of the effective mass m_{eff} of the superfluid mechanical modes. Thus, below we focus our attention on the computation of m_{eff} [59].

In continuum mechanics the effective mass can be obtained by shrinking the whole system to a reduction point $\vec{A}$ with effective mass m_{eff}. Then if the point $\vec{A}$ moves with velocity $v(\vec{A})$, it should possess the same kinetic energy E_k as the original system, i.e. $m_{\mathrm{eff}} = \frac{2E_k}{v_A^2}$. The kinetic energy of the system is given by $\frac{1}{2}\int_V \rho\, v^2\,(\vec{r})\,\mathrm{d}^3\,(\vec{r})$, where ρ is its density. Computation of E_k, in turn, requires the knowledge of the velocity distribution $v\,(\vec{r})$. In the case of thin films of superfluid helium, the question of the distribution of superfluid velocity [70] and density [61] under the surface of the film is not a trivial problem. Thus, in order to circumvent this complexity, we utilize the equipartition theorem which allows us to substitute the kinetic energy E_k in the expression for m_{eff} with the van der Waals potential energy stored in the deformation of the film surface. Exploiting $U\,(z) = -\frac{\alpha_{\mathrm{vdw}}}{z^3}$ for the van der Waals potential energy per unit mass and taking advantage of the analogy with gravity waves [82] (see Sect. 1.10) for the potential energy we obtain:

$$\begin{aligned} E_{\mathrm{pot}} &= \rho \int_{\theta=0}^{2\pi}\int_{r=0}^{R}\left(\int_{z=0}^{d+\eta(r,\theta)} U\,(z)\,\mathrm{d}z\right) r\,\mathrm{d}r\,\mathrm{d}\theta - \rho \int_{\theta=0}^{2\pi}\int_{r=0}^{R}\left(\int_{z=0}^{d} U\,(z)\,\mathrm{d}z\right) r\,\mathrm{d}r\,\mathrm{d}\theta \\ &= \rho \int_{0}^{2\pi}\int_{0}^{R}\left(\int_{d}^{d+\eta(r,\theta)} U\,(z)\,\mathrm{d}z\right) r\,\mathrm{d}r\,\mathrm{d}\theta. \end{aligned} \tag{2.5}$$

Here, $\eta(r,\theta)$ is the profile of the surface deformation of the superfluid film and d is its mean thickness.

In the limit when the amplitude of the surface oscillation is much smaller than the mean film thickness, i.e. $\eta \ll d$, we can Taylor expand $U\,(z)$ and compute the integral along the z-axis in Eq. (2.5). Then the expression for E_{pot} becomes:

$$E_{\mathrm{pot}} = 2\pi\rho \int_0^R \mathrm{d}r\, r \left(-\frac{\alpha_{\mathrm{vdw}}\,\eta\,(r,\theta)}{d^3} + \frac{3\,\alpha_{\mathrm{vdw}}\,\eta^2\,(r,\theta)}{2\,d^4}\right), \tag{2.6}$$

where α_{vdw} is the van der Waals coefficient for the WGM cavity material.

For the case of rotationally invariant third-sound modes, i.e. azimuthal mode order $m = 0$ (see Sect. 1.10) and $\eta(r,\theta) = \eta(r)$, and for free (volume-conserving)

Table 2.1 Scaling of superfluid experimental parameters with WGM cavity radius R, superfluid film thickness d and third-sound mode order ξ. Symbols – and † represent no dependence and a non-monotonous dependence respectively

	R	d	ξ
$m \propto$	R^2	d	–
$m_{\text{eff}} \propto$	R^4	d^{-1}	ξ^{-2}
$\Omega_M \propto$	R^{-1}	$d^{-3/2}$	ξ
$x_{\text{ZPF}} \propto$	$R^{-3/2}$	$d^{5/4}$	$\xi^{1/2}$
$g_0 \propto$	†	$d^{5/4}$	†

boundary conditions, i.e. $\int_0^R r\,\mathrm{d}r\,\eta(r) = 0$, we arrive at the expression for the effective mass of a point on the superfluid film surface at $r = R$:

$$m_{\text{eff}} = \left(\frac{\rho}{\rho_s}\right)\left(\frac{R}{d}\right)^2 \frac{1}{\xi^2} \times 2\pi\rho\, d\, \frac{\int_0^R r\,\eta^2(r)\,\mathrm{d}r}{\eta^2(R)}. \tag{2.7}$$

To obtain this, we also used $\Omega_M = \frac{\xi\, c_3}{R}$ for the third-sound mechanical frequency, where ξ is the Bessel mode frequency parameter, and $c_3 = \sqrt{3\frac{\rho_s}{\rho}\frac{\alpha_{\text{vdw}}}{d^3}}$ for the third-sound phase velocity (see Sect. 1.10).

The second multiplier in Eq. (2.7) can be recognized as the effective mass of a point on the boundary of a thin solid circular resonator of thickness d (for rotationally invariant modes) [83]. Therefore, while the effective mass of a point on a solid resonator scales as $R^2 d$ (like the real mass), for the case of a third-sound wave m_{eff} scales as R^4/d. This result is quite remarkable, as it implies that thicker and heavier films make lighter oscillators with larger amplitude of zero-point fluctuations (Eq. (2.3)). Moreover, the R^4 dependence of m_{eff} suggests that fabricating smaller third-sound resonators can be greatly beneficial for quantum optomechanical experiments with thin films of superfluid helium.

While an elaborated discussion of how such parameters as resonator dimensions, film thickness and third-sound mode order influence the optomechanical parameters of thin superfluid films is provided in Ref. [59], here we summarize the scaling of the superfluid experimental parameters with R, d, and ξ in Table 2.1.

2.3 Experimental Configuration

2.3.1 *Optical Cavity and Coupling*

In order to realize optomechanics with thin films of superfluid helium in practice, we developed an experimental configuration which allows an efficient interaction between the optical field and the mechanical motion of the film. This is achieved by immersing a high-quality optical cavity in a microscale droplet of superfluid helium. Both the emergence of superfluidity in helium and the quantum regime of optomechanical interactions require cryogenic conditions. Thus, our experimental set-up is contained inside an Oxford Instruments Triton closed-cycle helium-3 cryostat with a base temperature of 300 mK. All the experiments presented in this thesis were conducted with the Oxford helium-3 cryostat. We have also installed a BlueFors LD-250 dilution refrigerator with a base temperature of 10 mK. This refrigerator will be used for the next generation of the superfluid optomechanics experiments [84].

A whispering-gallery-mode optical microcavity (typically either a microtoroid or a microdisk, see Sect. 1.9; historically, a WGM microtoroid was used for the first generation of the superfluid optomechanics experiments reported in this thesis; further experiments with wedged microdisks that enable higher evanescent overlap with the superfluid helium film [59] are beyond the scope of this thesis and are reported in [84]) is enclosed within a sealed sample chamber attached to the coldest plate of the cryostat. Laser light at telecom wavelength (1550 nm band) is delivered inside the sample chamber and, following an interaction with the optical cavity, brought back to the room-temperature environment via a single-mode optical fibre. The signal detection occurs within a room-temperature optical set-up that will be described in Sect. 2.4.2 (see Fig. 2.7a). Inside the sample chamber light evanescently couples into the WGM cavity via a tapered region of the optical fibre. Depending on the experimental needs, either under-, critically-, or overcoupled optical regimes can be realized. This is achieved by mounting the optical cavity upon Attocube linear nanopositioners, which allow the distance between the tapered region of the fibre and the cavity to be adjusted. This enables different regimes of the optical coupling. The tapered fibre is mounted onto an in-house manufactured taper holder made of the same material as the optical fibre (silica) in order to match the thermal contraction/expansion during the cryostat cool-down and warm-up cycles. A set of windows at the bottom of the cryostat, reaching all the way to the sample chamber, allow imaging of the optical cavity-fibre configuration to be achieved through microscopes mounted just outside of the cryostat vacuum can. This imaging allows the cavity–fibre distance—and, hence, the optical coupling—to be tuned in a controlled manner.

In normal operation of the cryostat the pulse-tube cooler (PTC) runs continuously to maintain the temperature of the 40 K and 3 K cryogenic stages. However, the PTC causes vibrations which are transmitted from the top of the cryostat down to the sample chamber, hindering stabilization of the cavity–fibre separation. To prevent this, we designed on-chip stabilization silica beams (Fig. 2.3c) which are developed around the optical cavity (Fig. 2.3b) during its photolithographic fabrication process. One beam is placed on either side of the cavity (Fig. 2.3a) such that the tapered fibre is supported by the beams. This makes the relative chip–fibre vibration common-mode and allows an excellent suppression of the relative motion between the taper and cavity, enabling a high-precision stabilization of the cavity-fibre separation (similar precision is achieved in Ref. [85]). This technique allows us to indefinitely maintain the desired level of the optical coupling even while the PTC is in operation. In order to

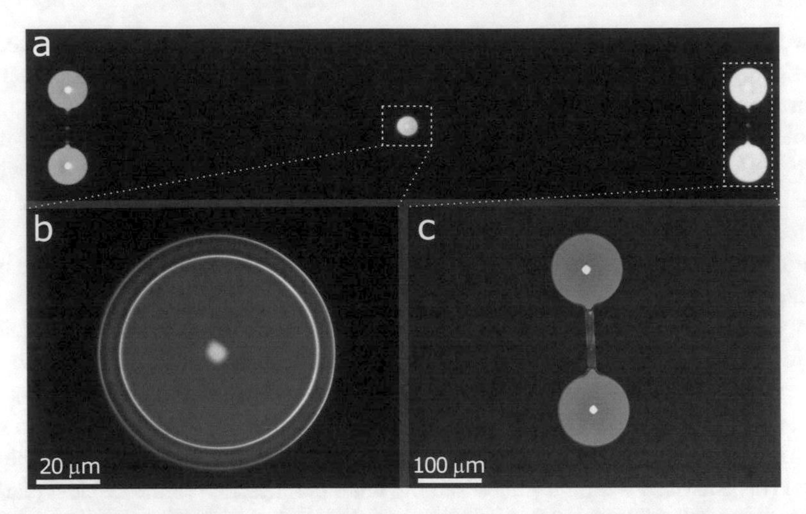

Fig. 2.3 **a** Optical microscope image of a microtoroid fabricated in-house and used in our experiments. The microtoroid is seen in the centre and its zoom-in is shown in (**b**). Support beams for tapered fibre stabilization, fabricated during the same photolithography flow as the microtoroid, are placed on either side of the cavity. **c** Zoom-in of a fibre support beam. Both support beams have been thinned down to ∼200 nm to minimize optical scattering losses. Microscope images were supplied by C. G. Baker

minimize light scattering losses caused by the strong optical evanescent field leaking out of the fibre, we made the silica beams only 5 μm wide and thinned them from 2 μm down to about 200 nm to make their thickness much smaller than the light wavelength. While the PTC does not affect mechanical motion of the optical cavity, it can perturb motion of the superfluid in the sample chamber. Therefore, during the measurements on the superfluid we switch the PTC off.

2.3.2 Superfluid Film Formation

In order to initiate the light–superfluid interactions, helium is first injected into the sample chamber in its gaseous state (Fig. 2.5 (left)), i.e. above the superfluid phase transition temperature (typical injection temperature is around 3 K). Helium-4 gas is introduced into the sample chamber via a stainless steel capillary tube connecting the sample chamber with an external helium reservoir kept at room temperature. Once a desired amount of helium is injected into the sample chamber, the capillary link between the chamber and the ambient environment is shut via a vacuum-tight valve. As can be seen from the phase diagram in Fig. 2.4, at low pressures, such as those used in our experiments, helium-4 transitions from its gaseous phase directly into the superfluid state (He II in Fig. 2.4), avoiding the normal fluid phase (He I in Fig. 2.4). Upon cooling our system down across the superfluid transition temperature, helium-4 gas in the sample chamber condenses into a thin superfluid film (Fig. 2.5 (right)). A combination of ultralow viscosity of the superfluid and attractive van der Waals

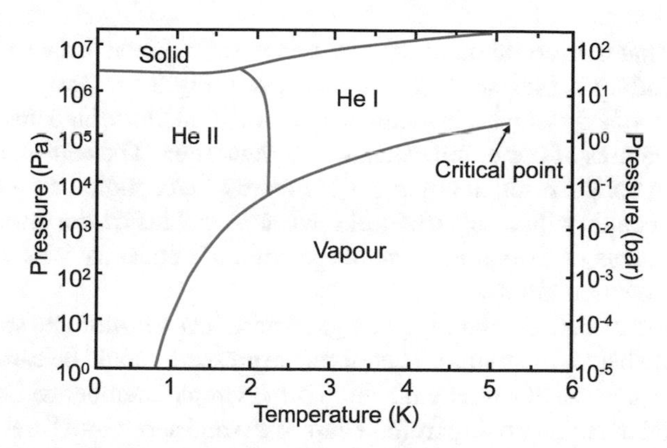

Fig. 2.4 Helium-4 phase diagram. Depending on temperature and pressure conditions helium-4 can exist in either gaseous (vapour), solid, or fluid phases. The latter can be either a normal fluid phase (He I) which has viscosity as any other classical fluid, or a superfluid phase (He II) which is inviscid. At low gas pressures used in our experiments (typically below 70 mTorr $\approx$ 0.1 mbar at $\sim$3 K) helium-4 transitions from the vapour phase directly into its superfluid state. The phase diagram is reproduced from the data of Ref. [75]

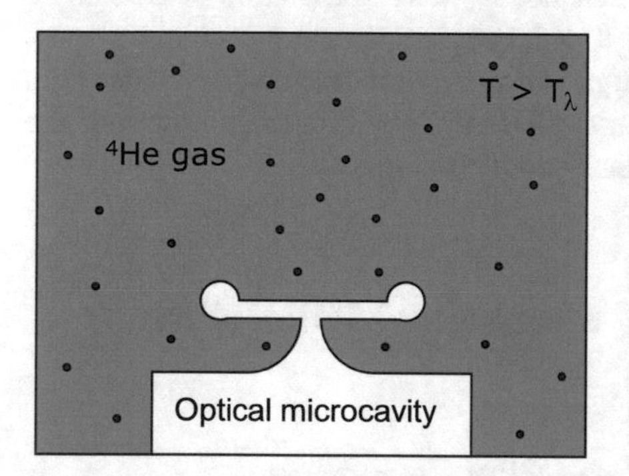

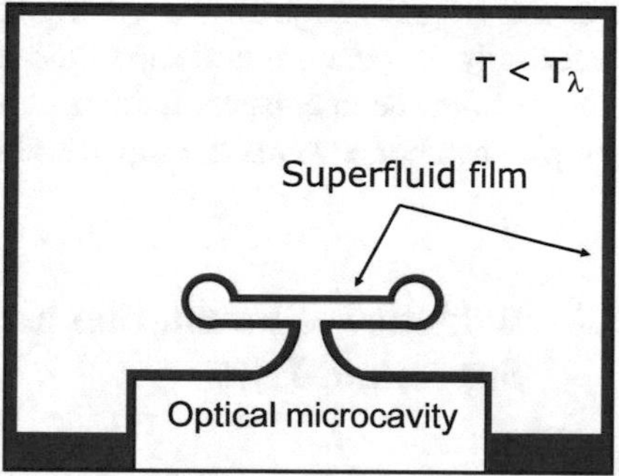

Fig. 2.5 (Left) A sketch of the sample chamber containing an optical microcavity and filled with helium-4 gas above the superfluid phase transition temperature. (Right) Upon cooling down and crossing the superfluid transition temperature, helium-4 gas condenses into a thin film of superfluid helium. Ultralow viscosity and attractive van der Waals forces coat the entire inner surface of the chamber with the superfluid film, including the optical cavity. The self-assembling nature of the film implies that an optical cavity of any type and shape can be exploited in our optomechanical configuration. Thicker film at the bottom of the sample chamber in the sketch on the right represents a pool of superfluid helium formed at certain pressures of the injected helium-4 gas (see text for details). T_λ stands for the superfluid phase transition temperature ("lambda-point"). Optical microcavity, helium atoms, and superfluid film are drawn not to scale

forces enable natural formation of the film on the entire inner surface of the sample chamber, including the surface of the optical cavity (Fig. 2.5 (right)). We would like to emphasize the self-assembling nature of the superfluid film which manifests as one of the great features of our optomechanical configuration. The self-assembling film implies that an optical resonator of any type (microsphere, photonic crystal etc.) will be uniformly coated with a superfluid film when placed inside the sample chamber. This feature allows us to explore a broad spectrum of optical cavities which we can exploit in our configuration.

The pressure of the injected helium-4 gas determines the ultimate superfluid film thickness at the base temperature. Therefore, the pressure should be carefully chosen to optimize the amount of superfluid helium in the sample chamber while minimizing the detrimental heat link arising from helium-4 gas and superfluid "creeping" up the injection capillary tube. The requirement of a constant chemical potential on the entire film surface and a balance between gravity and the attractive van der Waals forces, holding the film on the walls of the chamber, determine the maximal film thickness achievable at a given height above the sample chamber bottom [86]. At pressures of the injected helium-4 gas exceeding those corresponding to the maximal allowed film thickness, a pool of superfluid helium forms at the bottom of the sample chamber (Fig. 2.5 (right)). We would like to point out that the sample chamber must necessarily be vacuum- and superfluid-tight in order to avoid leakage of superfluid helium from the enclosure. Indium wire, used to seal any detachable parts of the sample chamber, ensures the superfluid leak-proof conditions.

2.4 Evidence of Formation and Mechanical Motion of the Superfluid Film

2.4.1 Observation of the Superfluid Film Formation

The optomechanical interaction between light[3] confined to the periphery of the WGM microtoroid and helium in the sample chamber (see Sect. 2.2.1) provides a toolkit to observe the superfluid film formation experimentally. We start off by coupling light into the microtoroid at around 3 K, which is above the superfluid phase transition temperature. Light transmitted through the optical cavity is probed via a 10% tap-off placed right after the microtoroid (Fig. 2.7a). A typical Lorentzian-type optical resonance is observed in the transmission spectrum of the cavity (Fig. 2.6a (yellow trace)). This is an indication that no film is formed on the surface of the resonator yet. However, as soon as the temperature is decreased just below the superfluid transition point, unstable oscillations emerge on the blue-detuned side of the optical resonance (Fig. 2.6a (blue trace)). While the temperature is being decreased further,

[3]This section incorporates a fraction of the work published by Springer Nature Publishing AG: G. I. Harris, D. L. McAuslan, E. Sheridan, **Y. Sachkou**, C. Baker, W. P. Bowen. Laser cooling and control of excitations in superfluid helium. *Nature Physics* 12, 788–793, 2016.

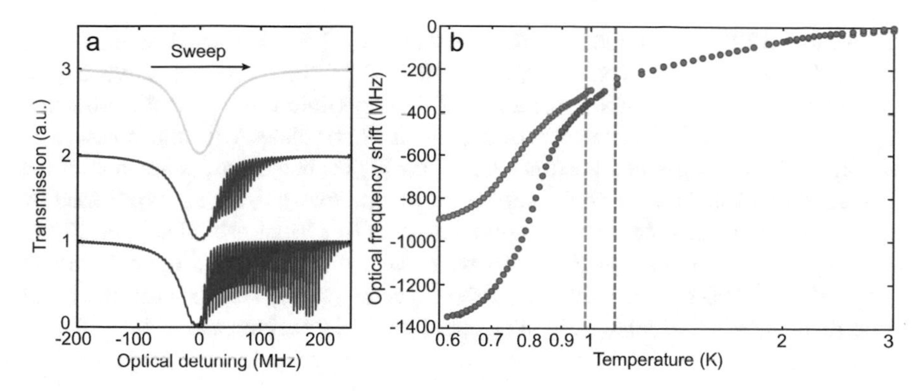

Fig. 2.6 **a** Spectra of optical transmission through the microtoroid above (yellow trace; 3 K) and below (blue (1 K) and red (0.6 K) traces) the superfluid transition temperature. Oscillations on the blue-detuned side of the blue and red traces are indicative of fluctuations of the superfluid film on the surface of the microtoroid. The "sweep" arrow indicates the direction of the laser frequency scanning in order to track the microtoroid optical resonance. The traces are offset vertically for clarity. **b** Frequency shift of the microtoroid optical resonance as a function of temperature for two different helium gas pressures. Orange circles—45 mTorr, blue circles—91 mTorr (both measured with the cryostat at 2.8 K). Different helium gas pressures in the sample chamber result in a different superfluid film thickness causing different dispersive shift of the microtoroid optical resonance. Vertical dashed bars indicate superfluid phase transition temperatures for each of the pressures. Figure is reproduced from Ref. [57] with modifications

the oscillations become more pronounced (Fig. 2.6a (red trace)). This behaviour is typical of optomechanical parametric instability. An optical field detuned to the blue side of the cavity resonance sheds energy into the mechanical oscillator via optical dynamical backaction, amplifying the mechanical motion [87]. This amplification results in a sharp rise of the motional amplitude of the oscillator, leading to a modification of the microtoroid effective radius. Consequently, this shifts the WGM resonance frequency, resulting in an alternating coupling of the laser light into the microtoroid which is imprinted as oscillations on the blue side of the transmission spectrum. We ascribe this oscillatory mechanism to fluctuations of the superfluid film formed on the surface of the microtoroid. An extremely low optical power of just 40 nW was enough to observe these oscillations. This power is one hundred times lower than would be required to excite similar oscillations of a microtoroid mechanical mode [87]. Moreover, the temperature of the unstable oscillations onset exactly agrees with the superfluid phase transition temperature predicted from the phase diagram for the helium pressure in our experiment (Fig. 2.4). All of these factors indicate that the oscillations in the optical transmission spectrum are caused by the fluctuations of the superfluid film on the microtoroid surface.

However, we go further and probe the superfluid film formation by measuring frequency of one of the microtoroid optical modes as a function of the decreasing cryostat temperature (Fig. 2.6b). Using Eq. (2.1), we computed that a 1 nm thick superfluid film should shift the microtoroid optical resonance frequency by -290 MHz nm^{-1}. Prior to the film formation, i.e. above the superfluid transition temperature, the optical frequency shift does not depend on the pressure of the helium gas and occurs purely due to the thermal expansion of the resonator and the thermorefractive effect [88] (see Sect. 1.9). However, upon formation of the superfluid film below the transition temperature the optical resonance frequency shift correlates with the pressure of the injected helium gas and becomes larger for higher helium pressures due to the formation of a thicker film on the surface of the microtoroid. This is confirmed by increasing the injected helium pressure from 45 mTorr (orange circles in Fig. 2.6b) to 91 mTorr (blue circles in Fig. 2.6b) which resulted in an extra optical shift of −457 MHz at 600 mK. Dividing it by the expected optical shift per nanometer of superfluid yields $457/290 = 1.58$ nm of the film thickness increase.

2.4.2 *Read-Out of the Superfluid Film Mechanical Motion*

2.4.2.1 Optical Set-Up

Once we confirmed the presence of the superfluid film on the microtoroid surface and its impact on the WGM optical mode, we then attempt to detect mechanical motion of the film. To achieve this, we again leverage the optomechanical interaction between light confined inside the microtoroid and the superfluid film. As elaborated in Sect. 2.2.1, thickness fluctuations of the film induced by the third-sound wave in the vicinity of the WGM perturb its optical evanescent field (Fig. 2.7b). This perturbation is resolved via the homodyne detection scheme within an all-fibre interferometer shown in Fig. 2.7a.

The superfluid film motion is quite fragile and prone to any external unwanted noise. Therefore, care has to be taken to ensure stable operation of the interferometer.

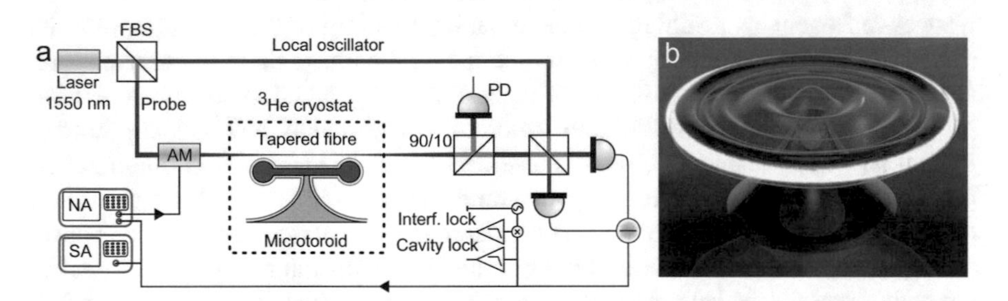

Fig. 2.7 **a** Shot-noise limited all-fibre homodyne detection scheme for resolution of mechanical motion of superfluid thin film. A WGM optical microtoroid is placed inside a sealed sample chamber attached to the coldest plate of the cryostat. Light is evanescently coupled into the microtoroid via a tapered optical fibre. FBS—fibre beam splitter; AM—amplitude modulator; PD—photodetector; NA/SA—network/spectrum analyzer. The 90/10 tap-off right after the microtoroid probes the optical transmission through the cavity. **b** Artist's representation of a microtoroid coated with thin film of superfluid helium. Glow at the microtoroid periphery represents light confined inside the cavity. Ripples on the surface of the superfluid film—third sound—alter the effective optical path length for light inside the microtoroid. This leads to a shift of the WGM optical resonance frequency which is resolved via the homodyne detection scheme in (**a**). Rendering was supplied by C. G. Baker

To do this, we implemented two locking loops, one of which locks the relative phase between the interferometer arms whilst the second ensures that the laser frequency is locked to the microtoroid optical resonance (respectively “Interf. lock” and “Cavity lock” in Fig. 2.7a). In order to fix the relative phase angle between the probe and the local oscillator interferometer arms (Fig. 2.7a), we apply an amplitude modulation with a frequency of 190 MHz before the microtoroid. Then the AC component of the generated photocurrent at the modulation frequency is mixed down with a 190 MHz signal derived directly from the signal generator. The result of the mix-down produces a phase dependent error signal for the interferometer which is passed through a filter and an amplifier and fed into a piezoelectric fibre stretcher which stabilizes the relative phase between the interferometer arms. In our configuration, the DC component of the photocurrent also provides a dispersive error signal for the cavity. After passing the DC component through a filter and an amplifier, it is fed into the laser port for the frequency stabilization, thus locking the laser frequency to the optical resonance of the microtoroid. The cavity lock error signal is amplified by the local oscillator, allowing a stable operation of our system even with nanowatts of optical power in the probe arm of the interferometer.

2.4.2.2 Characterization of the Superfluid Film Mechanical Motion

In order to observe and characterize mechanical motion of the superfluid film we lock the laser frequency to an optical resonance of the cavity. Light circulating at the periphery of the microtoroid generates heat via optical absorption in the silica. The heating induces a photoconvective superfluid flow through the thermomechanical (fountain) effect whereby superfluids tend to flow towards a localized heat source [42, 58] (for details see Sect. 1.5 and Chap. 3). We found that, as a consequence of the spatial overlap between the generated flow field and the third-sound modes, it is possible to coherently drive mechanical motion of the superfluid film through an amplitude modulation of the light confined inside the microtoroid. The optical amplitude modulation drives the third-sound modes via the photoconvective actuation. In order to detect the modulated light we mix it with a local oscillator (interferometer arm carrying light with intensity much higher than in the probe arm) derived from the same laser source and perform balanced homodyne detection [78] (Fig. 2.7a). While the laser light is coupled exactly to the WGM optical resonance (i.e. not detuned), the superfluid motion is directly imprinted onto the phase quadrature of the optical field (the optical amplitude quadrature contains no information about the motion of the mechanical oscillator in this case) [1, 2, 78]. The homodyne detection enables an ultrasensitive read-out of the optical phase which is directly proportional to the motional amplitude of the superfluid film [1, 2]. We then use spectrum and network analyzers to measure the detected signal.

The network analyzer allows us to drive some particular mechanical mode with a frequency of interest and to measure its instantaneous response. Sending a network analyzer drive signal to the amplitude modulator (Fig. 2.7a) enables sweeping of the optical modulation frequency. Recording the superfluid response with the network analyzer, we acquired a characteristic mode spectrum shown in Fig. 2.8a.

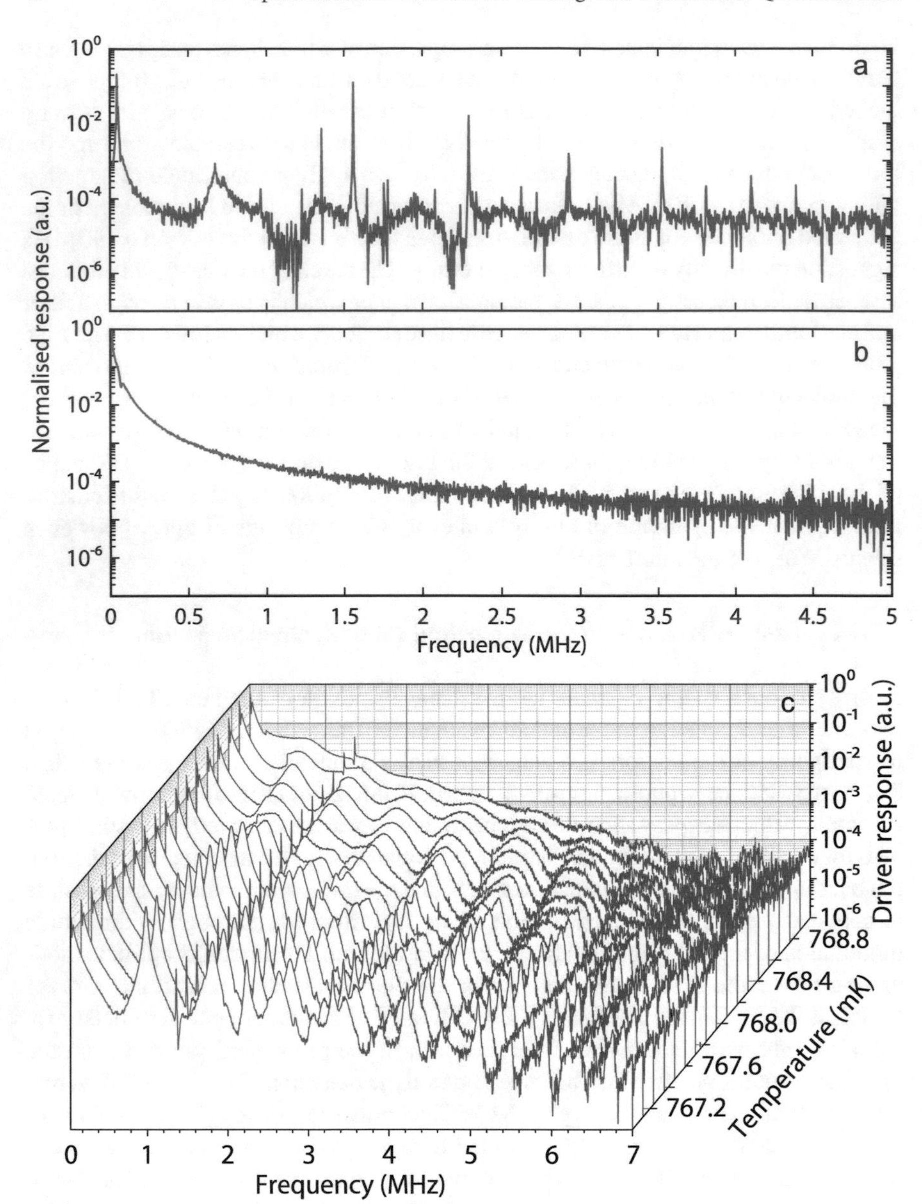
a
b
c
Normalised response (a.u.)
Frequency (MHz)
Driven response (a.u.)
Temperature (mK)
Frequency (MHz)

◀**Fig. 2.8** **a** Spectrum of superfluid mechanical modes (third-sound) measured with the network analyzer at 900 mK. Each distinct peak in the spectrum corresponds to a particular third-sound mode. **b** Network analysis measurement of the mechanical response above the superfluid transition temperature (1.1 K). No third-sound modes are observed, meaning absence of the superfluid film on the microtoroid surface. The large kink at low frequencies is caused by the thermo-optic response of silica. Once the superfluid layer is formed below the transition temperature, the kink vanishes, as can be seen in (**a**) [88, 89]. **c** Emergence of third-sound modes in the mechanical response of the superfluid film measured with the network analyzer. The spectra are recorded in a narrow temperature window of 2 mK (from 767 to 769 mK), with a step of 0.2 mK, right across the superfluid phase transition temperature. Each emergent peak in the spectra corresponds to a particular third-sound mode. The sharp peak at 1.35 MHz corresponds to a microtoroid mechanical mode and stays unchanged throughout the temperature variation. This data was acquired for a helium-4 gas pressure of 19 mTorr at 2.8 K compared to the typical 69 mTorr at 2.8 K throughout most of the work in this chapter. This explains the lower superfluid transition temperature (see Fig. 2.4). Figure is reproduced from Ref. [57] with modifications

The spectrum features a range of mechanical modes with frequencies from 10 kHz to 5 MHz which are consistent with expected frequencies of third-sound modes confined within a domain with the dimensions of the microtoroid. In order to confirm that the observed peaks in the spectrum do indeed correspond to the superfluid third-sound modes, and not to some other oscillatory mechanism, we probed the spectrum both below and above the superfluid transition temperature (1 K at 69 mTorr of helium gas pressure—see phase diagram in Fig. 2.4). The mentioned spectrum in Fig. 2.8a was recorded at 0.9 K, while the one in Fig. 2.8b—at 1.1 K—displays no modes. Since the peaks in the spectrum appear only below the superfluid phase transition temperature, this is indicative that they represent mechanical resonances in the superfluid film. We would also like to note that the kink at low frequencies in Fig. 2.8b is attributed to the thermo-optic response of silica which the microtoroid is made of. However, as soon as a layer of superfluid helium forms on the microtoroid surface below the superfluid transition temperature, the thermo-optic response reduces significantly and the kink vanishes (Fig. 2.8a) due to an improved microtoroid thermalization provided by the superfluid layer.

In order to further investigate the emergence of the third-sound spectrum, we performed network analysis measurements of the superfluid film mechanical response in a narrow temperature window of 2 mK right across the superfluid phase transition (Fig. 2.8c). Traces were recorded with a step of 0.2 mK. Figure 2.8c shows how peaks corresponding to third-sound modes rapidly arise as the temperature is swept across the phase transition. A sharp peak at 1.35 MHz corresponds to a microtoroid mechanical mode and remains unchanged as the temperature is varied. Comparison of this microtoroid mode frequency with other peaks in the spectrum at different temperatures should serve as another confirmation of the superfluid nature of the emergent modes in Fig. 2.8c. Figure 2.9a shows the temperature dependence of two different superfluid mechanical modes as well as of the 1.35 MHz microtoroid mode. The superfluid mode frequencies are observed to rapidly drop with decreasing temperature. This is caused by the film thickening, which is a consequence of the increased condensation of helium atoms into the superfluid film at lower temperatures.

Thicker film results in a weaker van wer Waals restoring force, hence, "softer" mechanical oscillators, and thus lower superfluid mechanical frequencies at lower temperatures. In contrast, the frequency of the 1.35 MHz microtoroid mechanical mode, as expected, experiences a negligible variation of less than 0.1% across the same temperature range (Fig. 2.9a).

Frequencies of the superfluid modes plateau at low temperatures as all of the helium atoms have been condensed into the superfluid film. This provides a degree of freedom to tune the third-sound mode frequencies on-demand by simply varying the helium gas pressure prior to the superfluid film formation. We, therefore, measured frequency of a particular superfluid mechanical mode as a function of pressure of helium gas injected in the sample chamber (Fig. 2.9b). As predicted, the frequency of the superfluid mode was observed to decrease at the base temperature at higher pressure values. This can be explained by a larger number of helium atoms injected in the sample chamber at higher pressures, which results in a thicker film at the base temperature and, hence, lower frequency of the superfluid mechanical mode. Also the superfluid phase transition temperature was noted to be higher with increasing gas pressures, which is consistent with the helium-4 phase diagram [75] (Fig. 2.4).

Finally, we would like to note that, while measuring the superfluid mechanical response and gradually increasing the drive strength, we observed that at sufficiently high drive powers superfluid modes exhibit Duffing nonlinearity [65, 90]. In order to ensure that all measurements were carried out in the linear regime, we kept the drive powers well below the one around which the nonlinear features were observed.

2.5 Real-Time Tracking of Superfluid Thermomechanical Motion

Superfluid mechanical motion[4] (third sound in two-dimensional superfluids) can be driven either by some external driving force (as described above in Sect. 2.4.2) or by the intrinsic elementary excitations in a quantum fluid. Dynamics of the elementary excitations and interactions between them dictate both micro- and macroscopic properties of a superfluid. One form of the intrinsic elementary excitations in superfluids manifests as thermal fluctuations, such as phonons and rotons [42]. In order to understand the microscopic properties of superfluids, it is crucial to be able to track, and possibly manipulate, their thermally driven excitations in real time, i.e. as their dynamics occur.

A thermally driven mechanical mode of a superfluid can be treated as an oscillator coupled to an external thermal bath at temperature T. The equipartition theorem [91] gives the root-mean-square (RMS) motional amplitude of the oscillator as $\delta x = \sqrt{k_B T/k}$, with k_B being the Boltzmann constant and k the spring constant of the

[4] This section incorporates a part of the work published by Springer Nature Publishing AG: G. I. Harris, D. L. McAuslan, E. Sheridan, **Y. Sachkou**, C. Baker, W. P. Bowen. Laser cooling and control of excitations in superfluid helium. *Nature Physics* 12, 788–793, 2016.

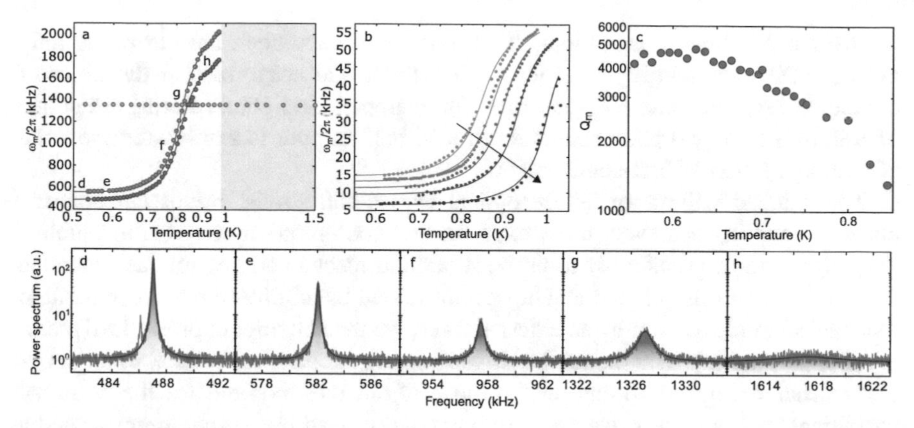

Fig. 2.9 **a** Mechanical frequencies of two particular superfluid modes (orange and blue) and a microtoroid mechanical mode (green) as a function of the cryostat temperature. At low temperatures frequencies of the superfluid modes plateau as all helium atoms condensed into the film. Kinks in the superfluid modes temperature dependence around 900 mK are dictated by the nonlinearity in both the pressure at which the superfluid phase transition occurs (see Fig. 2.4) and third-sound velocity as a function of temperature. The grey-shaded area represents the temperature range above the superfluid transition where all helium is in its gaseous phase. **b** Temperature dependence of the frequency of a particular third-sound mode at different pressures of helium-4 gas injected in the sample chamber. The black arrow indicates the direction of the pressure increase. As expected from the helium-4 phase diagram (Fig. 2.4), higher helium pressure results in a higher superfluid transition temperature. The final mode frequency at low temperatures is lower for higher helium pressures as the film is thicker and, thus, "softer". The measured third-sound mode is different from those presented in (**a**). **c** Mechanical quality factor of the 482 kHz third-sound mode as a function of the cryostat temperature. Due to low SNR above 850 mK it was not possible to accurately measure the quality factor beyond that temperature. **d**–**h** Thermomechanical spectra of the 482 kHz third-sound mode at cryostat temperatures of 530, 700, 800, 850, and 900 mK (from left to right). The corresponding temperatures of the spectra are denoted in (**a**). Figure is reproduced from Ref. [57] with modifications

oscillator. Dissipative processes determine that the oscillator's mechanical motion decorrelates over a characteristic time scale of $1/\Gamma_m$, where Γ_m is the oscillator decay rate. In order to probe and control the instantaneous thermodynamic fluctuations of the oscillator in real time it is essential to resolve the oscillation amplitude with precision better than δx within a time frame shorter than $1/\Gamma_m$. Such a measurement rate would allow to track the thermally driven trajectory of the oscillator in phase space. In practice, a real-time measurement implies an ability to resolve the peak of the thermally driven motion of the oscillator above the measurement noise floor in the power spectral density of the measurement. In other words, a signal-to-noise ratio (SNR) greater than one should be achieved faster than $1/\Gamma_m$. A rigorous theoretical justification of why the SNR should equal exactly unity is beyond the scope of this thesis, however it is provided in the Supplementary Information of Ref. [57].

To date, many experimental techniques have been developed to probe intrinsic thermal dynamics of phonon elementary excitations in superfluid helium [42, 68, 69]. However, their measurement rates have proved to be slow compared to characteristic dissipation of phonon excitations. Therefore, they have been able to probe only averaged thermodynamic properties of superfluids and not to resolve dynamics of the elementary excitations in real time. For example, the light scattering technique of Ref. [69] averaged photocounts for around half an hour to resolve the spectrum of thermal motion of first-sound waves.

The reduced volume of WGM optical modes and strong evanescent optomechanical coupling achieved in our experimental configuration combine to enable a very high read-out sensitivity to probe superfluid mechanical motion, as described in Sect. 2.4.2 of this chapter. Having demonstrated the ability to resolve superfluid mechanical modes driven by an external force, we then attempt to probe the dynamics of superfluid intrinsic thermally–driven phonon excitations. We test whether our optomechanical approach can enable the read-out rate required for the real-time measurement. To do this, we perform homodyne-based phase measurement and a subsequent spectral analysis on a high-quality third-sound mode at 482 kHz. The emergence of a Lorentzian-type thermomechanical noise peak attributed to the mode with decreasing cryostat temperature is shown in Fig. 2.9d–h. The decay rate of this third-sound mode is expressed in terms of its mechanical quality factor shown in Fig. 2.9c. As the cryostat cools down, the quality factor increases, consistent with previous observations reported in Refs. [65, 92]. The decay rate reaches its minimum of $\Gamma_m/2\pi = 106$ Hz at 530 mK. The noise peak reaches SNR of 21 dB at this temperature (Fig. 2.9d), which significantly exceeds unity, thus enabling measurement of the superfluid mechanical motion in real time.

To demonstrate real-time measurement, we tracked the superfluid motion encoded onto the homodyne photocurrent over a period of 5 s. We then applied filters of varying length to the recorded signal in order to simulate different measurement times. The best estimate of the phase-space mode position (amplitude of the X-quadrature) was obtained with a single-sided exponential filter of length $1/\Gamma_m$. Filters of shorter length were then applied to the data in order to acquire a set of measurements of the oscillator position in phase space. Next, all these measurements were compared to the best estimate to obtain the measurement variance. In order to determine the minimum measurement time required to resolve the thermomechanical motion, the ratio of the measurement variance, V, to the thermal variance, V_T (best estimate), was obtained as a function of the measurement duration τ and shown in Fig. 2.10a. As can be seen from this figure, measuring for a shorter time results in a larger variance. We define a measurement as being real-time if the measurement variance is less than half of the thermal variance. The superfluid motion is resolvable for measurement times as low as 121 μs, a factor of 11 times shorter than $1/\Gamma_m$ (1.3 ms), and thus sufficient to track the superfluid motion in real time.

It is illustrative to depict motional quadratures of an oscillator in phase space. We monitored a thermal trajectory of the 482 kHz superfluid mode (blue line in Fig. 2.10b) in real time from a series of 1.3 ms-long measurements, i.e. duration of each of the measurements is equal to $1/\Gamma_m$. Each orange point represents the oscillator position measured within 1.3 ms, and therefore the orange area is the thermal cloud of the third-sound mode. The blue circle represents the imprecision of an individual measurement, computed as the standard deviation of the shot-noise

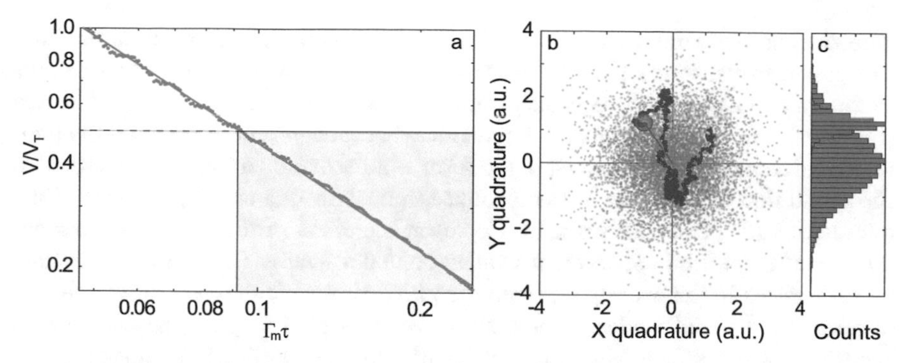

Fig. 2.10 **a** The ratio of the measurement variance V to the thermal variance V_T as a function of the measurement time represented in terms of its ratio to the oscillator decorrelation time. The grey-shaded area corresponds to the regime when the measurement variance is smaller than half of the oscillator thermal variance, i.e. where the measurement is real-time. **b** Thermal trajectory of the 482 kHz third-sound mode in phase space (blue line), tracked in real time from a series of 1.3 ms-long measurements. Each orange point represents the position of the mode, measured within 1.3 ms, i.e. $1/\Gamma_m$. The entire orange area is the thermal cloud of the oscillator. The blue circle is the measurement imprecision of an individual measurement, computed as the standard deviation of the shot noise within the time of the measurement. **c** Normalized histograms of the measurement imprecision (blue) and the thermomechanical noise of the superfluid third-sound mode (orange). Uncertainties of both the superfluid position and the measurement are distributed normally. Measurement imprecision is 4 times smaller than the superfluid thermomechanical noise. Figure is reproduced from Ref. [57] with modifications

within the measurement time. When the measurement time τ increases, the area of the blue circle diminishes, implying that the measurement imprecision decreases. The measurement is defined as real-time if the area of the measurement imprecision (blue circle) is smaller than the thermal cloud of the oscillator (orange area). As can be seen from the histogram in Fig. 2.10c, we tracked the superfluid mechanical motion with the measurement imprecision (blue histogram) a factor of 4 below the superfluid thermomechanical noise (orange histogram).

2.6 Cooling and Heating of Superfluid Mechanical Modes

Our optomechanical approach[5] to probing superfluid thermal motion allows not only to track superfluid mechanical modes but also to manipulate their behaviour. In order to demonstrate this, we cool and heat phonon excitations of the superfluid film. This is achieved by red-detuning the laser frequency from the WGM cavity resonance (in contrast, real-time tracking of superfluid thermal motion was carried out on resonance; also note that in regular optomechanics heating only happens on the

[5]This section incorporates a part of the work published by Springer Nature Publishing AG: G. I. Harris, D. L. McAuslan, E. Sheridan, **Y. Sachkou**, C. Baker, W. P. Bowen. Laser cooling and control of excitations in superfluid helium. *Nature Physics* 12, 788–793, 2016.

blue side) and utilizing dynamical backaction from both optical radiation pressure and photothermal forces [93]. By varying the detuning Δ we observed alterations in the third-sound modes resonance frequency and linewidth, i.e. dissipation rate (Fig. 2.11). We ascertained that the dynamical backaction is dominated by the photothermal forces, which manifest in the form of an actuation of the photoconvective superfluid flow in response to the localized optical heating at the periphery of the microtoroid [58] (see also Sect. 3). However, the effect of the dynamical backaction on different third-sound modes varies quite drastically. This might be dictated by differences in the overlap between the photoconvective flow and flow fields of different third-sound modes. Remarkably, even modes of very similar frequencies experience very different photothermal forces. Thus, some superfluid modes appear split in the spectrum, as can be seen in Fig. 2.11. The splitting is attributed to the dispersive interaction between sound modes and quantized vortices residing in the superfluid film. A detailed discussion of the phonon-vortex interactions is presented in Chaps. 4 and 5 of this thesis. However, what is relevant to the current chapter is that the higher-frequency peak of these near-degenerate mode pairs was consistently observed to experience stronger dynamical backaction than the lower frequency one (Fig. 2.11). One possible explanation of this is a cylindrical symmetry breaking caused by a miniscule departure of the microtoroid from perfect circularity during the microfabrication process. Then the symmetry is broken not only for the third-sound modes, but also for the photoconvective flow field. Therefore, superfluid modes with symmetry matching the flow field might be driven, whereas those with the opposing symmetry might be damped.

The spatial overlap between the photoconvective flow and flow fields of the superfluid modes determines not only the strength of the photothermal response, but also its sign. Whether a third-sound mode is cooled or heated depends on the interplay between the radiation pressure force and the sign of the photothermal response. Examples of cooling and heating of two distinct third-sound modes are shown in Figs. 2.11a and 2.11b respectively. The 552.5 kHz mode exhibits linewidth broadening (cooling) from $\Gamma_0/2\pi = 115$ Hz at zero detuning ($\Delta = 0$) to $\Gamma_\Delta/2\pi = 464$ Hz at a detuning of $\Delta = -0.58\kappa$, where κ is the intrinsic linewidth. The photothermal response of this mode has the same qualitative characteristics as radiation pressure, namely broadening of the mechanical mode accompanied by spring softening when red-detuned from the optical cavity. In contrast to the 552.5 kHz mode, the 482 kHz mode exhibits linewidth narrowing (heating) and spring stiffening with increasing detuning Δ (Fig. 2.11b). This is indicative of the negative sign of the photothermal response which opposes the direction of the radiation pressure force. The linewidth of this mode narrowed from $\Gamma_0/2\pi = 137$ Hz at $\Delta = 0$ to $\Gamma_\Delta/2\pi = 49$ Hz at $\Delta = -0.58\kappa$. From the change in dissipation rates (linewidth) we computed that the 552.5 kHz mode was cooled by a factor of $\frac{T_\Delta}{T_0} = \frac{\Gamma_0}{\Gamma_\Delta} = 0.25$, whereas the 482 kHz mode was heated by a factor of $\frac{T_\Delta}{T_0} = 2.8$.

We would like to point out that the characteristic time of the photothermal response in our system is $\tau \approx 600$ ns. Such an outstandingly fast response is enabled by the very high thermal conductivity of superfluid helium [42].

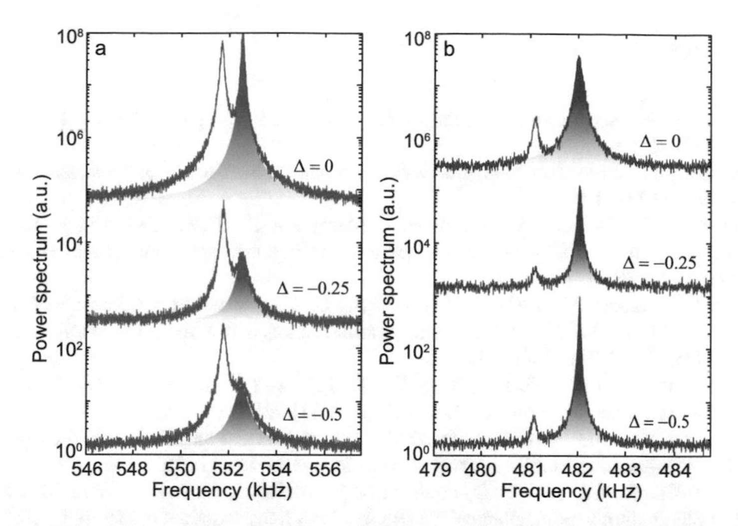

Fig. 2.11 Optomechanical cooling (**a**) and heating (**b**) of two distinct superfluid third-sound modes. Both cooling and heating were achieved by red-detuning the laser frequency from the WGM cavity resonance and utilizing dynamical backaction from both optical radiation pressure and photothermal forces. The interplay between the radiation pressure and the sign of the photothermal response determines whether a mode is cooled or heated. The shaded higher frequency peak in (**a**) exhibits linewidth *broadening* with increasing detuning Δ, resulting in a decreasing area under the peak and, thus, mode cooling. In contrast, the shaded higher frequency peak in (**b**) exhibits linewidth *narrowing* with increasing detuning Δ and hence mode heating. The spectral splitting of both modes is ascribed to the dispersive interaction between third-sound modes and quantized vortices in the film. Traces in (**a**) and (**b**) are offset for clarity. All measurements were conducted with 200 nW of input optical power. Figure is reproduced from Ref. [57] with modifications

2.7 Conclusion

We presented an approach to cavity optomechanics based on the interaction between optical field confined within a high-quality whispering-gallery-mode resonator and mechanical excitations on the surface of a thin film of superfluid helium. We showed that low effective mass of the films results in large amplitude of zero-point motion and, hence, large single-photon optomechanical coupling and cooperativity. Small optical mode volume and strong optomechanical coupling results in an enhanced sensitivity of the optical field to perturbations on the resonator surface presented by the superfluid film. This allowed us to resolve and track superfluid thermomechanical excitations in real time, which, to our best knowledge, is the first demonstration of such capability. Furthermore, we showed laser cooling and heating of superfluid mechanical modes. Our optomechanical platform with thin films of superfluid helium offers a great potential for both cavity optomechanics and investigation of two-dimensional quantum fluids.

References

1. Aspelmeyer M, Kippenberg TJ, Marquardt F (2014) Cavity optomechanics. Rev Mod Phys 86(4):1391–1452
2. Kippenberg TJ, Vahala KJ (2008) Cavity optomechanics: back-action at the mesoscale. Science 321(5893):1172–1176
3. Thompson JD, Zwickl BM, Jayich AM, Marquardt F, Girvin SM, Harris JGE (2008) Strong dispersive coupling of a high-finesse cavity to a micromechanical membrane. Nature 452(7183):72–75
4. Jayich AM, Sankey JC, Børkje K, Lee D, Yang C, Underwood M, Childress L, Petrenko A, Girvin SM, Harris JGE (2012) Cryogenic optomechanics with a Si_3N_4 membrane and classical laser noise. New J Phys 14(11):115018
5. Purdy TP, Peterson RW, Yu P-L, Regal CA (2012) Cavity optomechanics with Si_3N_4 membranes at cryogenic temperatures. New J Phys 14(11):115021
6. Arcizet O, Cohadon P-F, Briant T, Pinard M, Heidmann A (2006) Radiation-pressure cooling and optomechanical instability of a micromirror. Nature 444(7115):71–74
7. Favero I, Stapfner S, Hunger D, Paulitschke P, Reichel J, Lorenz H, Weig EM, Karrai K (2009) Fluctuating nanomechanical system in a high finesse optical microcavity. Opt Express 17(15):12813–12820
8. Carmon T, Rokhsari H, Yang L, Kippenberg TJ, Vahala KJ (2005) Temporal behavior of radiation-pressure-induced vibrations of an optical microcavity phonon mode. Phys Rev Lett 94(22):223902
9. Tomes M, Carmon T (2009) Photonic micro-electromechanical systems vibrating at X-band (11-GHz) rates. Phys Rev Lett 102(11):113601
10. Baker C, Hease W, Nguyen D-T, Andronico A, Ducci S, Leo G, Favero I (2014) Photoelastic coupling in gallium arsenide optomechanical disk resonators. Opt Express 22(12):14072–14086
11. Eichenfield M, Chan J, Camacho RM, Vahala KJ, Painter O (2009) Optomechanical crystals. Nature 462(7269):78–82
12. Teufel JD, Donner T, Li D, Harlow JW, Allman MS, Cicak K, Sirois AJ, Whittaker JD, Lehnert KW, Simmonds RW (2011) Sideband cooling of micromechanical motion to the quantum ground state. Nature 475(7356):359–363
13. Chan J, Alegre TPM, Safavi-Naeini AH, Hill JT, Krause A, Gröblacher S, Aspelmeyer M, Painter O (2011) Laser cooling of a nanomechanical oscillator into its quantum ground state. Nature 478(7367):89–92
14. Wollman EE, Lei CU, Weinstein AJ, Suh J, Kronwald A, Marquardt F, Clerk AA, Schwab KC (2015) Quantum squeezing of motion in a mechanical resonator. Science 349(6251):952–955
15. Pirkkalainen J-M, Damskägg E, Brandt M, Massel F, Sillanpää MA (2015) Squeezing of quantum noise of motion in a micromechanical resonator. Phys Rev Lett 115(24):243601
16. Chu Y, Kharel P, Yoon T, Frunzio L, Rakich PT, Schoelkopf RJ (2018) Creation and control of multi-phonon Fock states in a bulk acoustic-wave resonator. Nature 563(7733):666
17. Ockeloen-Korppi CF, Damskägg E, Pirkkalainen J-M, Asjad M, Clerk AA, Massel F, Woolley MJ, Sillanpää MA (2018) Stabilized entanglement of massive mechanical oscillators. Nature 556(7702):478
18. Riedinger R, Wallucks A, Marinković I, Löschnauer C, Aspelmeyer M, Hong S, Gröblacher S (2018) Remote quantum entanglement between two micromechanical oscillators. Nature 556(7702):473
19. Metcalfe M (2014) Applications of cavity optomechanics. Appl Phys Rev 1(3):031105
20. Mamin HJ, Rugar D (2001) Sub-attonewton force detection at millikelvin temperatures. Appl Phys Lett 79(20):3358–3360
21. Forstner S, Prams S, Knittel J, van Ooijen ED, Swaim JD, Harris GI, Szorkovszky A, Bowen WP, Rubinsztein-Dunlop H (2012) Cavity optomechanical magnetometer. Phys Rev Lett 108(12):120801

22. Chaste J, Eichler A, Moser J, Ceballos G, Rurali R, Bachtold A (2012) A nanomechanical mass sensor with yoctogram resolution. Nat Nanotechnol 7(5):301–304
23. Lucklum R, Hauptmann P (2006) Acoustic microsensors-the challenge behind microgravimetry. Anal Bioanal Chem 384(3):667–682
24. Ramos D, Mertens J, Calleja M, Tamayo J (2008) Phototermal self-excitation of nanomechanical resonators in liquids. Appl Phys Lett 92(17):173108
25. Yu W, Jiang WC, Lin Q, Lu T (2014) Coherent optomechanical oscillation of a silica microsphere in an aqueous environment. Opt Express 22(18):21421–21426
26. Fong KY, Poot M, Tang HX (2015) Nano-optomechanical resonators in microfluidics. Nano Lett 15(9):6116–6120
27. Gil-Santos E, Baker C, Nguyen DT, Hease W, Gomez C, Lemaître A, Ducci S, Leo G, Favero I (2015) High-frequency nano-optomechanical disk resonators in liquids. Nat Nanotechnol 10(9):810–816
28. Linden J, Thyssen A, Oesterschulze E (2014) Suspended plate microresonators with high quality factor for the operation in liquids. Appl Phys Lett 104(19):191906
29. Burg TP, Godin M, Knudsen SM, Shen W, Carlson G, Foster JS, Babcock K, Manalis SR (2007) Weighing of biomolecules, single cells and single nanoparticles in fluid. Nature 446(7139):1066–1069
30. Bahl G, Kim KH, Lee W, Liu J, Fan X, Carmon T (2013) Brillouin cavity optomechanics with microfluidic devices. Nat Commun 4:1994
31. Han K, Kim KH, Kim J, Lee W, Liu J, Fan X, Carmon T, Bahl G (2014) Fabrication and testing of microfluidic optomechanical oscillators. J Vis Exp (87)
32. Zhu K, Han K, Carmon T, Fan X, Bahl G (2014) Opto-acoustic sensing of fluids and bioparticles with optomechanofluidic resonators. Eur Phys J Spec Top 223(10):1937–1947
33. Hyun Kim K, Bahl G, Lee W, Liu J, Tomes M, Fan X, Carmon T (2013) Cavity optomechanics on a microfluidic resonator with water and viscous liquids. Light Sci Appl 2(11):e110
34. Han K, Zhu K, Bahl G (2014) Opto-mechano-fluidic viscometer. Appl Phys Lett 105(1):014103
35. Maayani S, Martin LL, Carmon T (2016) Water-walled microfluidics for high-optical finesse cavities. Nat Commun 7:10435
36. Maayani S, Martin LL, Kaminski S, Carmon T (2016) Cavity optocapillaries. Optica 3(5):552–555
37. Giorgini A, Avino S, Malara P, De Natale P, Gagliardi G (2017) Fundamental limits in high-Q droplet microresonators. Sci Rep 7:41997
38. Dahan R, Martin LL, Carmon T (2016) Droplet optomechanics. Optica 3(2):175–178
39. Giorgini A, Avino S, Malara P, De Natale P, Yannai M, Carmon T, Gagliardi G (2018) Stimulated Brillouin cavity optomechanics in liquid droplets. Phys Rev Lett 120(7):073902
40. Labrador-Páez L, Soler-Carracedo K, Hernández-Rodríguez M, Martín IR, Carmon T, Martin LL (2017) Liquid whispering-gallery-mode resonator as a humidity sensor. Opt Express 25(2):1165–1172
41. De Lorenzo LA, Schwab KC (2014) Superfluid optomechanics: coupling of a superfluid to a superconducting condensate. New J Phys 16(11):113020
42. Tilley DR, Tilley J (1990) Superfluidity and superconductivity. CRC Press
43. Seidel GM, Lanou RE, Yao W (2002) Rayleigh scattering in rare-gas liquids. Nucl Instrum Methods Phys Res Sect A Accel Spectrom Detect Assoc Equip 489(1):189–194
44. De Lorenzo LA, Schwab KC (2017) Ultra-high Q acoustic resonance in superfluid ^{4}He. J Low Temp Phys 186(3–4):233–240
45. De Lorenzo LA (2016) Optomechanics with superfluid Helium-4. PhD thesis, California Institute of Technology
46. Singh S, De Lorenzo LA, Pikovski I, Schwab KC (2017) Detecting continuous gravitational waves with superfluid ^{4}He. New J Phys 19(7):073023
47. Souris F, Christiani H, Davis JP (2017) Tuning a 3D microwave cavity via superfluid helium at millikelvin temperatures. Appl Phys Lett 111(17):172601
48. Rojas X, Davis JP (2015) Superfluid nanomechanical resonator for quantum nanofluidics. Phys Rev B 91(2):024503

49. Souris F, Rojas X, Kim PH, Davis JP (2017) Ultralow-dissipation superfluid micromechanical resonator. Phys Rev Appl 7(4):044008
50. Kashkanova AD, Shkarin AB, Brown CD, Flowers-Jacobs NE, Childress L, Hoch SW, Hohmann L, Ott K, Reichel J, Harris JGE (2017) Superfluid Brillouin optomechanics. Nat Phys 13(1):74–79
51. Kashkanova AD, Shkarin AB, Brown CD, Flowers-Jacobs NE, Childress L, Hoch SW, Hohmann L, Ott K, Reichel J, Harris JGE (2017) Optomechanics in superfluid helium coupled to a fiber-based cavity. J Opt 19(3):034001
52. Shkarin AB, Kashkanova AD, Brown CD, Garcia S, Ott K, Reichel J, Harris JGE (2017) Quantum optomechanics in a liquid. arXiv: 1709:02794
53. Childress L, Schmidt MP, Kashkanova AD, Brown CD, Harris GI, Aiello A, Marquardt F, Harris JGE (2017) Cavity optomechanics in a levitated helium drop. Phys Rev A 96(6):063842
54. Fong KY, Jin D, Poot M, Bruch A, Tang HX (2018) Phonon coupling between a nanomechanical resonator and a quantum fluid. arXiv: 1803.07552
55. Bradley DI, George R, Guénault AM, Haley RP, Kafanov S, Noble MT, Pashkin YA, Pickett GR, Poole M, Prance JR, Sarsby M, Schanen R, Tsepelin V, Wilcox T, Zmeev DE (2017) Operating nanobeams in a quantum fluid. Sci Rep 7(1):4876
56. Guénault AM, Guthrie A, Haley RP, Kafanov S, Pashkin YA, Pickett GR, Tsepelin V, Zmeev DE, Collin E, Gazizulin R, Maillet O, Arrayás M, Trueba JL (2018) Driving nanomechanical resonators by phonon flux in superfluid ^{4}He. arXiv: 1810.10129
57. Harris GI, McAuslan DL, Sheridan E, Sachkou Y, Baker C, Bowen WP (2016) Laser cooling and control of excitations in superfluid helium. Nat Phys 12(8):788–793
58. McAuslan DL, Harris GI, Baker C, Sachkou Y, He X, Sheridan E, Bowen WP (2016) Microphotonic forces from superfluid flow. Phys Rev X 6(2):021012
59. Baker CG, Harris GI, McAuslan DL, Sachkou Y, He X, Bowen WP (2016) Theoretical framework for thin film superfluid optomechanics: towards the quantum regime. New J Phys 18(12):123025
60. Sachkou YP, Baker CG, Harris GI, Stockdale OR, Forstner S, Reeves MT, He X, McAuslan DL, Bradley AS, Davis MJ, Bowen WP (2019) Coherent vortex dynamics in a strongly interacting superfluid on a silicon chip. Science 366(6472):1480–1485
61. Scholtz JH, McLean EO, Rudnick I (1974) Third sound and the healing length of He II in films as thin as 2.1 atomic layers. Phys Rev Lett 32(4):147–151
62. Yang G, Fragner A, Koolstra G, Ocola L, Czaplewski DA, Schoelkopf RJ, Schuster DI (2016) Coupling an ensemble of electrons on superfluid helium to a superconducting circuit. Phys Rev X 6(1):011031
63. Nasyedkin K, Byeon H, Zhang L, Beysengulov NR, Milem J, Hemmerle S, Loloee R, Pollanen J (2018) Unconventional field-effect transistor composed of electrons floating on liquid helium. J Phys Condens Matter 30(46):465501
64. Bishop DJ, Reppy JD (1978) Study of the superfluid transition in two-dimensional He 4 films. Phys Rev Lett 40(26):1727
65. Hoffmann JA, Penanen K, Davis JC, Packard RE (2004) Measurements of attenuation of third sound: evidence of trapped vorticity in thick films of superfluid ^{4}He. J Low Temp Phys 135(3–4):177–202
66. Ellis FM, Luo H (1989) Observation of the persistent-current splitting of a third-sound resonator. Phys Rev B 39(4):2703–2706
67. Barenghi CF, Skrbek L, Sreenivasan KR (2014) Introduction to quantum turbulence. Proc Natl Acad Sci 111:4647–4652
68. Bramwell ST, Keimer B (2014) Neutron scattering from quantum condensed matter. Nat Mater 13(8):763–767
69. Pike ER, Vaughan JM, Vinen WF (1970) Brillouin scattering from superfluid He-4. J Phys Part C Solid State Phys 3(2):L40
70. Atkins KR (1959) Third and fourth sound in liquid Helium II. Phys Rev 113(4):962–965
71. Everitt CWF, Atkins KR, Denenstein A (1962) Detection of third sound in liquid helium films. Phys Rev Lett 8(4):161–163

72. Schechter AMR, Simmonds RW, Packard RE, Davis JC (1998) Observation of 'third sound' in superfluid He-3. Nature 396(6711):554–557
73. Anetsberger G, Arcizet O, Unterreithmeier QP, Rivière R, Schliesser A, Weig EM, Kotthaus JP, Kippenberg TJ (2009) Near-field cavity optomechanics with nanomechanical oscillators. Nat Phys 5(12):909–914
74. Anetsberger G, Weig EM, Kotthaus JP, Kippenberg TJ (2011) Cavity optomechanics and cooling nanomechanical oscillators using microresonator enhanced evanescent near-field coupling. C R Phys 12(9–10):800–816
75. Donnelly RJ, Barenghi CF (1998) The observed properties of liquid helium at the saturated vapor pressure. J Phys Chem Ref Data 27(6):1217–1274
76. Rosencher E, Vinter B (2002) Optoelectronics. Cambridge University Press
77. Kippenberg TJ, Spillane SM, Armani DK, Vahala KJ (2003) Fabrication and coupling to planar high-Q silica disk microcavities. Appl Phys Lett 83(4):797–799
78. Bowen WP, Milburn GJ (2015) Quantum optomechanics. CRC Press
79. Jiang X, Lin Q, Rosenberg J, Vahala K, Painter O (2009) High-Q double-disk microcavities for cavity optomechanics. Opt Express 17(23):20911–20919
80. Sabisky ES, Anderson CH (1973) Verification of the Lifshitz theory of the van der Waals potential using liquid-helium films. Phys Rev A 7(2):790–806
81. Keller WE (1970) Thickness of the static and the moving saturated He II film. Phys Rev Lett 24(11):569–573
82. Phillips OM (1969) The dynamics of the upper ocean. Cambridge University Press, London
83. Wang J, Ren Z, Nguyen CT (2004) 1.156-GHz self-aligned vibrating micromechanical disk resonator. IEEE Trans Ultrason Ferroelect Freq Control 51(12):1607–1628
84. He X, Harris GI, Baker CG, Sawadsky A, Sfendla YL, Sachkou YP, Forstner S, Bowen WP (2020) Strong optical coupling through superfluid Brillouin lasing. Nat Phys 16(4):417–421
85. Barclay PE, Srinivasan K, Painter O, Lev B, Mabuchi H (2006) Integration of fiber-coupled high-Q SiNx microdisks with atom chips. Appl Phys Lett 89(13)
86. Enss C, Hunklinger S (2005) Low-temperature physics. Springer
87. Harris GI, Andersen UL, Knittel J, Bowen WP (2012) Feedback-enhanced sensitivity in optomechanics: surpassing the parametric instability barrier. Phys Rev A 85(6):061802
88. Arcizet O, Rivière R, Schliesser A, Anetsberger G, Kippenberg TJ (2009) Cryogenic properties of optomechanical silica microcavities. Phys Rev A 80(2):021803
89. Riviere R, Arcizet O, Schliesser A, Kippenberg TJ (2013) Evanescent straight tapered-fiber coupling of ultra-high Q optomechanical micro-resonators in a low-vibration helium-4 exchange-gas cryostat. Rev Sci Instrum 84(4):043108
90. Rips S, Kiffner M, Wilson-Rae I, Hartmann MJ (2012) Steady-state negative Wigner functions of nonlinear nanomechanical oscillators. New J Phys 14:023042
91. Jeans J (1982) An introduction to the kinetic theory of gases. Cambridge University Press, Cambridge Cambridgeshire
92. Penanen K, Packard RE (2002) A model for third sound attenuation in thick ^{4}He films. J Low Temp Phys 128(1–2):25–35
93. Restrepo J, Gabelli J, Ciuti C, Favero I (2011) Classical and quantum theory of photothermal cavity cooling of a mechanical oscillator. C R Phys 12(9–10):860–870

Chapter 3
Light-Mediated Control of Superfluid Flow

Optical forces are widely used in microphotonic systems. By driving the motion of mechanical elements, they enable applications ranging from precision sensing and metrology to quantum information and fundamental science. To date, the primary approaches to optical forcing exploit either the direct radiation pressure from light [1], or photothermal forces [2, 3] where optical heating causes mechanical stress and subsequent deformation of a mechanical element. Photothermal forces have the advantage of allowing significantly stronger actuation capabilities, but to-date have proved incompatible with the cryogenic conditions required to reach the quantum regime. In this work we demonstrate a new approach to optical forcing that allows strong microphotonic forces to be achieved in cryogenic conditions. The approach utilises, for the first time in a microphotonic context, the well-known fountain effect in superfluid helium [4], whereby superfluids flow convectively towards a heat source. In our case, the heat is generated by optical absorption in the vicinity of a mechanical element. A force is exerted when the fluid reaches this element and imparts momentum onto it. We experimentally achieve, within a cryogenic environment, microphotonic forces that are an order of magnitude stronger than their radiation pressure counterparts. As a demonstration of the utility of our technique, we use the superfluid photoconvective force to feedback cool a mechanical mode of a microtoroidal resonator to temperatures as low as 137 mK. Depending on geometry of a microphotonic system, photoconvective flow can be utilised to exert a wide range of forces, including, for instance, torques and compression, providing a versatile tool for cryogenic actuation of microphotonic systems.[1]

[1]This chapter is based on the work published by the American Physical Society: D. L. McAuslan, G. I. Harris, C. Baker, **Y. Sachkou**, X. He, E. Sheridan, and W. P. Bowen. Microphotonic Forces from Superfluid Flow. *Physical Review X*, 6(2):021012, 2016.

Y. Sachkou, *Probing Two-Dimensional Quantum Fluids with Cavity Optomechanics*, Springer Theses, https://doi.org/10.1007/978-3-030-52766-2_3

3.1 Introduction

Optical forces are extensively exploited in various micro- and nanophotonic systems, ranging from photonic circuits [5, 6] and ultracold matter [7] to optical tweezers [8, 9] and biophysics [10, 11]. In cavity optomechanics, in particular, optical forces actuate microscale mechanical elements, enabling a wide variety of applications from ultraprecise mass/force/field/spin sensors [12–15] to quantum and classical information systems [16]. Strong actuation is essential for an efficient manipulation of mechanical oscillators in the quantum regime. For instance, strong forces are particularly important for a wide class of quantum optomechanics protocols where measurement and feedback are used to ground-state cool [17] or prepare nonclassical states of a mechanical oscillator, such as squeezed [18–20], phonon number [21] and superposition states [22].

To date, two main approaches to optical forcing utilize either the direct radiation pressure from light [1], or photothermal forces [2, 3], where optical absorption and heating induce mechanical stress and subsequent deformation of a mechanical element. While radiation pressure and static gradient optical forces have proved instrumental for demonstration of, for instance, ground state cooling [1], ponderomotive squeezing [23], coherent state-swapping [24], and all-optical routing [25], photothermal forces are known to permit the strongest optical actuation capabilities [26]. For instance, an optomechanical system in which cooling via dynamical backaction was demonstrated for the first time was indeed based on the photothermal interaction [2]. Moreover, photothermal forces have been shown to enable efficient optomechanical cooling of a low-frequency mechanical oscillator to its quantum ground state in a bad-cavity limit [26]. However, to ensure strong actuation, photothermal systems are typically restricted to materials with a large thermal expansion coefficient and substantial optical absorption. Furthermore, such materials are precluded from cryogenic operation due to a several-orders-of-magnitude reduction in the thermal expansion coefficient when cooled to cryogenic temperatures typical for quantum optomechanics experiments.

In this work we demonstrate a novel, photoconvective, approach to microphotonic forcing based on the well-known superfluid *fountain effect* [4], whereby an optically induced heat source generates a convective superfluid flow that collides with the mechanical oscillator, thus transferring momentum from the fluid to the oscillator. The superfluid photoconvective forces presented in this work allow strong optical actuation to be achieved in cryogenic conditions. Furthermore, due to the presence of superflow, superfluid helium has the highest thermal conductivity of any known material. This enables strong thermal anchoring of the mechanical element, significantly alleviating challenges associated with parasitic heating of the mechanical oscillator due to the presence of light, which is one of the critical factors affecting the performance of cryogenic quantum optomechanics experiments [27].

Unlike other forces that are available in cryogenic conditions, superflow provides a mechanism to perform remote actuation, whereby the actuator is far removed from the mechanical element and forces are applied by the flow field that it generates.

This could be used to further minimize heating effects when performing optical actuation and also allows new actuation schemes, such as torques at microscale [28]. Moreover, owing to the ability of superfluid currents to persist for many hours without perceptible decay [29]—unique property of superfluids—superfluid flow offers a prospect to apply persistent forces on micromechanical elements long after the optical field has decayed, which could be used for nonvolatile optical logical elements such as routers [30] and memories [31].

3.2 Photoconvective Forces from Superfluid Flow

3.2.1 Radial Force

In order to investigate superfluid convective microphotonic forcing, we consider the configuration shown in Fig. 3.1a. A silica microtoroidal whispering-gallery-mode (WGM) optical resonator is supported by a silicon pedestal on a silicon chip (see Sect. 1.9). A thin film (typically <5 nm) of superfluid helium covers the entire surface of the resonator and the chip. Laser light, evanescently coupled into the resonator, is confined to the outer boundary of the microtoroid. Optical absorption of the laser field in silica creates a localized heat source at the periphery of the resonator (red glow in Fig. 3.1a). Heat-caused temperature increase triggers superfluid helium flow (blue arrows) towards the heat source via the superfluid thermomechanical ('fountain') effect [4]. Reaching the heat source, at low pressures, superfluid helium converts directly into gas and evaporates from the surface of the resonator (red arrows). Departing helium atoms exert a force onto the microtoroid. However, the total force experienced by the resonator also contains a contribution from the radiation pressure, whereby photons confined inside the WGM cavity impart an outwards radial force on the resonator boundary due to momentum exchange during the total internal reflection [32]. Here, in order to compare the magnitude of the superfluid photoconvective force to its radiation pressure counterpart, we evaluate a net inward radial force exerted by the evaporating helium atoms:

$$F_{\text{radial}} = -\frac{\mathrm{d}(m v_{\text{radial}})}{\mathrm{d}t}, \tag{3.1}$$

where m is the mass of vaporized helium and v_{radial} is the radial velocity.

Thus, in order to calculate the above radial force, we have to estimate the mass of helium evaporated per unit of time $\dot{m}$, i.e. the superfluid mass flow rate caused by the optical absorption, and compute the radial velocity of evaporating helium atoms. But before embarking on this, we need to make an important assumption that all the heat flow in the system occurs via superfluid convective heat transfer and not through conduction in the silica, i.e. heat generated by optical absorption is entirely dissipated solely by the superfluid. This assumption is important for the steady state

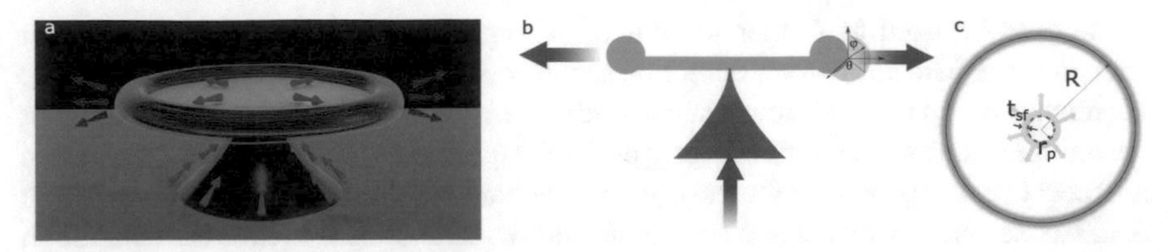

Fig. 3.1 **a** Artist's representation of our experimental configuration utilized to investigate superfluid microphotonic forcing. A microtoroidal whispering-gallery-mode optical cavity is coated with a thin film of superfluid helium. Laser light is confined to the outer boundary of the cavity. Optical absorption of light in silica generates heat at the periphery of the microtoroid (red glow). Temperature increase triggers superfluid flow (blue arrows) towards the localized heat source due to the superfluid thermomechanical effect. Reaching the heat source, superfluid evaporates from the surface of the microtoroid (red arrows) with helium atoms imparting momentum to the cavity. **b** Sketch of the cross-section of a microtoroidal optical cavity. The blue arrow represents superfluid flow which originates from the silicon pedestal and flows towards the area of localized heat at the periphery of the resonator. Velocity of superfluid flow on the pedestal is directed vertically upwards before becoming purely horizontal at the point of connection of the silicon pedestal with the silica microtoroid. Red arrows indicate evaporation of helium atoms from the microtoroid surface. **c** Sketch of the top view of a microtoroidal resonator of radius R (silica microtoroid is shown transparent for convenience). Boundary of the pedestal of radius r_p is shown in dashed line. Blue arrows indicate superflow at the level of the pedestal. The superfluid film thickness is labelled t_{sf}. Red glow indicates the optically generated heat. Figure is reproduced from Ref. [35] with modifications

condition described below and valid since the thermal conductivity κ_{sf} of superfluid helium is many orders of magnitude larger than that of silica κ_{SiO_2} at the temperatures used in our experiments [33, 34]:

$$\frac{\kappa_{sf}\, t_{sf}}{\kappa_{SiO_2}\, t_{SiO_2}} \gg 1. \tag{3.2}$$

Here t_{sf} and t_{SiO_2} are thickness' of superfluid film and silica microtoroid respectively. This assumption is also verified in our experiments, which confirm that once the superfluid film boils off at high optical powers, the resonator is no longer in thermal equilibrium with the cryogenic environment (see Sect. 3.3.2 and Fig. 3.4b).

3.2.1.1 Superfluid Mass Flow Rate

In order to theoretically investigate the recoil force on the microtoroid resulting from evaporation of helium atoms from its surface, we consider a steady state case where the heat load in the resonator is fully balanced out by the energy dissipation from the microtoroid surface. Heat power P_{abs} is generated through absorption of the circulating laser field in the silica at the microtoroid periphery. In order to balance out this heat load, energy should dissipate through normal fluid counterflow or evaporation of helium atoms. As we discussed in Sect. 1.10, the normal fluid component is viscously clamped to the surface of the microtoroid [36], which implies that normal fluid cannot flow out and, hence, cannot take heat away. This means that the energy dissipation occurs via evaporation of helium atoms (unlike in bulk superfluid systems where the energy dissipation is typically conducted via the normal fluid counterflow). Given the steady-state scenario, fluid cannot accumulate in the vicinity of the heat source on the microtoroid surface, which implies that the evaporation rate should be equal to the fluid influx from superfluid flow. For the optically generated heat power P_{abs}, the superfluid mass flow rate is then given by

$$\dot{m} = \frac{P_{abs}}{L - \langle \mu_{vdw} \rangle + \Delta H}. \tag{3.3}$$

Here, L is the latent heat of vaporization (in our experimental conditions $L \simeq$ 17.5 kJ/kg [37]) and $\Delta H = \int_{T_0}^{T_{evap}} c(T)\, \mathrm{d}T$ is the superfluid enthalpy change between the cryostat temperature T_0 and the evaporation temperature T_{evap}, where $c(T)$ is the specific heat capacity of liquid helium. However, the enthalpy change ΔH is small compared to the latent heat of vaporization L and, thus, can be neglected [37]. In Eq. (3.3), $\langle \mu_{vdw} \rangle$ represents the mean van der Waals potential energy of the superfluid film. We shall now focus our attention on how to compute this quantity.

3.2.1.2 Van der Waals Potential

The van der Waals interaction between helium atoms and atoms of the substrate material provides the restoring force for the superfluid film. Under the approximation of thin films, relevant to our experimental conditions, the van der Waals potential μ_{vdw} (which has the dimension of the energy per unit mass) at a distance d from the substrate is given by

$$\mu_{vdw} = -\frac{\alpha_{vdw}}{d^3}. \tag{3.4}$$

For silica—the material of our fabricated microtoroidal resonators—the van der Waals coefficient is equal to $\alpha_{vdw} = 2.65 \times 10^{-24}\ \mathrm{m^5 s^{-2}}$.

Since the normal fluid component has viscosity, we can consider that the first helium monolayer is viscously clamped to the substrate and is not involved in the heat transport [38]. Then the mean van der Waals potential of the flowing superfluid in our experimental case can be computed as

$$\langle \mu_{vdw} \rangle \simeq \frac{1}{2 - 0.35} \int_{x=0.35\mathrm{nm}}^{x=2\mathrm{nm}} \mu_{vdw}(x)\, \mathrm{d}x = -6.4\ \mathrm{kJ/kg}. \tag{3.5}$$

From the perspective of a helium atom, this is the energy required to escape the van der Waals potential. In Eq. (3.5) the approximate thickness of the first helium monolayer on silica resonator is equal to 0.35 nm, and the approximate superfluid film thickness is 2 nm. Hence, the integral in (3.5) is taken only over the flowing part

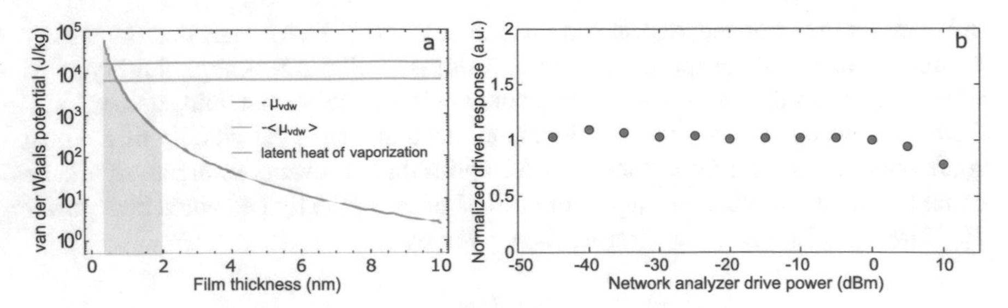

Fig. 3.2 **a** Absolute value of the van der Waals potential (3.4) as a function of film thickness d (blue). The blue-shaded area corresponds to the superfluid film thickness—the film thickness range relevant to our experiments—over which the mean van der Waals potential $\langle \mu_{\text{vdw}} \rangle$ is computed (orange). For comparison, the latent heat of vaporization for liquid helium at 500 mK is shown in green. **b** Normalized driven response of the microtoroid mechanical mode of interest as a function of the drive power in the presence of the superfluid film. The purpose of this measurement is to demonstrate that the superfluid photoconvective force linearly depends on the laser power in the optical cavity, as predicted by Eq. 3.10. To measure this driven response, we pass a laser beam (laser power is 1 μW) through an electro-optic modulator (EOM) and then send it straight into the microtoroid. We vary the drive power on the EOM and measure the driven response of the microtoroid mode using a network analyzer. The normalization—driven response divided by the drive power—is done such that a flat line corresponds to a linear response of the photoconvective force. The response is linear over a 50 dBm range, with the drop at the highest values of the drive power caused by a saturation of the EOM response. Figure is reproduced from Ref. [35] with minor modifications

of the film. Comparing the mean value of the van der Waals potential with values of the latent heat of vaporization L and the enthalpy change ΔH, one can see that the contribution of the van der Waals potential in Eq. (3.3) cannot be neglected. Indeed, $\langle \mu_{\text{vdw}} \rangle$ constitutes nearly 36% of the latent heat of vaporization for liquid helium. However, $\langle \mu_{\text{vdw}} \rangle$ is significantly greater than the change in enthalpy of the superfluid. Figure 3.2 shows the van der Waals potential as a function of the superfluid film thickness and the comparison of its mean value over the experimentally relevant film thickness range with the latent heat of vaporization.

Using Eq. (3.3) and taking a value for the optical heat power generated in the microtoroid through absorption to be $P_{\text{abs}} = 1\,\mu\text{W}$, we can now calculate the superfluid mass flow rate which yields $\dot{m} \approx 4.1 \times 10^{-11}$ kg/s.

3.2.1.3 Condensation of Helium Atoms on the Microtoroid Surface

In order to maintain the steady state condition, evaporated helium atoms should be replenished through the incoming superfluid flow. This can be achieved either via the fluid flow up the pedestal or through the condensation of helium atoms on the microtoroid surface after evaporation. Let us first consider the effect of the condensation and estimate its contribution in replenishing helium atoms.

Kinetic theory of gases gives the impingement rate r of gas molecules on a surface as [39]

$$r = \frac{\mathrm{P}}{\sqrt{2\pi k_{\mathrm{B}} T m_{\mathrm{He}}}}, \tag{3.6}$$

where P is the gas pressure, k_{B} the Boltzmann constant, T the temperature, and m_{He} the mass of a helium atom. The impingement rate is given in particles per unit area per unit time.

The upper bound of the mass flow rate due to condensation can be computed with an assumption that every single helium atom that impinges upon top and bottom surfaces of the microtoroid condenses on either of these surfaces. Then the mass flow rate is given by

$$\dot{m}_{\mathrm{condensation}} = r \frac{2\pi R^2}{N_A} M, \tag{3.7}$$

where R is the microtoroid major radius, $\mathrm{M} \simeq 4$ g/mol is the molar mass of helium, and N_A Avogadro's constant. We would like to point out that, from the energetic point of view, the condensation events counter-balance the evaporation counterparts, as, upon condensation, each atom releases its latent heat before taking away practically the same amount upon evaporation from the outer boundary of the resonator. Thus, the condensation events do not alter the results of Eq. (3.3).

At saturated vapour pressure $P \approx 10^{-2}$ Pa, the condensation mass flow rate computes to $\dot{m}_{\mathrm{condensation}} \approx 6.7 \times 10^{-13}$ kg/s. This upper bound of $\dot{m}_{\mathrm{condensation}}$ is nearly two orders of magnitude smaller than the superfluid mass flow rate $\dot{m}$ induced by the thermomechanical effect in our experiments. Thus, the effect of condensation of helium atoms on the microtoroid surface can safely be neglected. This leaves the only possible option for the incoming superfluid flow to maintain the steady state condition—to originate from the pedestal.

3.2.1.4 Computation of F_{radial}

Having determined the superfluid mass flow rate, we now have to compute the net radial velocity v_{radial} of helium atoms departing upon evaporation. In order to do this, we consider that from every point of the microtoroid periphery helium atoms evaporate isotropically in the outwards oriented half-space (see Fig. 3.1b). Then v_{radial} can be computed by integrating all contributions of this isotropic evaporation. Thus, the radial force (3.1) is given by

$$F_{\mathrm{radial}} = -\frac{\mathrm{d}(m v_{\mathrm{radial}})}{\mathrm{d}t} = \dot{m} v_{\mathrm{rms}} \frac{1}{\pi} \int_{-\pi/2}^{\pi/2} \cos(\theta)\, \mathrm{d}\theta \times \frac{1}{\pi} \int_{-\pi/2}^{\pi/2} \cos(\phi)\, \mathrm{d}\phi = \frac{4}{\pi^2} \dot{m} v_{\mathrm{rms}}, \tag{3.8}$$

where v_{rms} is a root-mean-square (rms) velocity of the departing atoms. As known from the kinetic theory of gases [39], v_{rms} is given by

$$v_{\mathrm{rms}} = \sqrt{\frac{3k_{\mathrm{B}}T_{\mathrm{evap}}}{m_{\mathrm{He}}}}. \tag{3.9}$$

Here, T_{evap} is the temperature of evaporated atoms, which in our experimental conditions is $\simeq$1 K.

The two integrals over the polar angle ϕ and the azimuthal angle θ in Eq. (3.8) represent projections of evaporated helium atoms' momenta onto the radial axis (see Fig. 3.1b).

Plugging expressions (3.3) for the superfluid mass flow rate and (3.9) for the rms-velocity of atoms in Eq. (3.8), we now arrive at the final result for the recoil radial force experienced by the microtoroid from the evaporation of helium atoms:

$$F_{\mathrm{radial}} = \frac{4}{\pi^2}\sqrt{\frac{3k_{\mathrm{B}}T_{\mathrm{evap}}}{m_{\mathrm{He}}}}\frac{P_{\mathrm{abs}}}{L - \langle\mu_{\mathrm{VDW}}\rangle}. \tag{3.10}$$

With the optically generated heat power of $P_{\mathrm{abs}} = 1\,\mu$W, the radial recoil force on the microtoroid computes to $F_{\mathrm{radial}} \simeq 1.3 \times 10^{-9}$ N.

It is important to note that the evaporation of helium atoms generates only the radial force, and does not exert a *net* force on the microtoroid in the vertical direction z. This is a consequence of the fact that the vaporization is isotropic along z axis and, thus, produces no *net* momentum change along z.

According to Eq. 3.10, F_{radial} should linearly depend on the injected laser power. And this is indeed consistent with our experimental observations, as can be seen in Fig. 3.2b.

3.2.1.5 Comparison of F_{radial} with Radiation Pressure Force

Once of the great advantages of the superfluid photoconvective forces is that, similar to photothermal forces [2], F_{radial} does not depend on the cavity finesse, as can be seen from Eq. (3.10). This means that the photoconvective forces can be efficient in configurations with a weak cavity or even no cavity at all.

We compare the radial superfluid photoconvective force—for the case when the incident light is fully absorbed—with its competitor—radiation pressure force, which is given by $F_{\mathrm{RP}} = P_{\mathrm{abs}}\mathcal{F}/c$, with $\mathcal{F}$ being the cavity finesse and c the speed of light. For configuration with no cavity, i.e. finesse $\mathcal{F} = 1$, and with the absorbed optical power $P_{\mathrm{abs}} = 1\,\mu$W, as in all previous calculations, the radiation pressure force computes to $F_{\mathrm{RP}} = 3 \times 10^{-15}$ N. For the superfluid evaporation temperature of 1 K, F_{radial} is $\simeq 4.3 \times 10^5$ times stronger than F_{RP}. This means that in our configuration a cavity with a finesse of approximately 430 000 would be required for the radiation pressure force to be comparable to the superfluid photoconvective force. However, the WGM cavity used in our experiments has a finesse of 53 000, which still makes the superfluid force one order of magnitude stronger than the one from the radiation pressure.

3.2.1.6 Efficiency of the Evaporation Process

It is intuitively clear that the more helium atoms evaporate from the surface per unit of time, the stronger recoil force they exert on the microtoroid. In order to estimate the efficiency of the recoil process, we calculate the number of helium atoms N evaporated in response to the heat generated by one absorbed photon:

$$N = \frac{\hbar\omega_0}{k_B\left(\mu + \frac{3}{2}T_{evap}\right)}. \tag{3.11}$$

Here, the numerator is the energy of the incoming photon, with ω_0 being the optical resonance frequency, and the denominator represents the energy required to evaporate one helium atom, with $\mu = 7.15$ K being the latent heat per atom (chemical potential) [40]. Taking the evaporation temperature of 1 K, yields approximately 1100 helium atoms evaporated per one absorbed photon. Multiplying N by the mass of a helium atom, we obtain the total mass of evaporated helium per photon: $m_{evap} \simeq 7.4 \times 10^{-24}$ kg. We can also estimate the fraction of the photon energy converted into the kinetic energy of the evaporated helium atoms. We do so by multiplying N by the kinetic energy of one atom and diving this by the photon energy. Using Eq. 3.9, we arrive at the following:

$$\eta = \frac{\frac{3}{2}T_{evap}}{\mu + \frac{3}{2}T_{evap}} \simeq 17\,\%. \tag{3.12}$$

3.2.2 *Vertical Force*

Even though there is no *net* vertical force on the microtoroid caused by the vaporization (see Sect. 3.2.1), there can be a vertical force exerted on the resonator by the superfluid flow arriving from the silicon pedestal. In response to the evaporation of helium atoms from the area of localized heat, superfluid continuously flows from the substrate to the microtoroid periphery via the pedestal. The induced superfluid flow, streaming up the silicon pillar, is purely vertical. Therefore, it reaches the silica resonator with momentum directed only vertically upwards. Upon arriving at the flat silica resonator, the flow becomes solely horizontal. The flow thereby exerts a vertical force on the microtoroid as the superfluid changes direction at the joint between the pedestal and the silica resonator. We would like to emphasize that this force is not a direct consequence of the evaporation of helium atoms, but rather caused by the momentum change of the macroscopic superflow. We can, therefore, in analogy with Eq. 3.8 define this vertical force F_z as

$$F_z = -\frac{d(mv_z)}{dt}. \tag{3.13}$$

Here v_z is velocity of the flow through the junction point between the pedestal and the microtoroid. This velocity can be computed in the following way.

We first obtain the superfluid volumetric flow rate $\dot{V}$ using Eq. 3.3:

$$\dot{V} = \frac{\dot{m}}{\rho} \simeq \frac{1}{\rho} \frac{P_{\text{abs}}}{L - \langle \mu_{\text{vdw}} \rangle} \tag{3.14}$$

Here $\rho = 145\ \text{kg/m}^3$ is the superfluid helium density. As we discussed in Sect. 3.2.1, all the incoming superfluid flow, streaming to replenish the evaporated helium, should come only through the pedestal. The flow velocity is the largest at the narrowest point of the pedestal, i.e. at the junction with the microtoroid. The velocity v_z through this point can be obtained as

$$v = \frac{\dot{V}}{2\pi r_{\text{p}} t_{\text{sf}}}. \tag{3.15}$$

Here t_{sf} and r_{p} are the superfluid film thickness and the pedestal radius respectively. The denominator $2\pi r_{\text{p}} t_{\text{sf}}$ represents the cross-sectional area of the pedestal together with the film. This is the area through which the helium influx occurs at the level of the pedestal (see Fig. 3.1c).

Plugging v_z from Eq. 3.15 into Eq. 3.13, we obtain the vertical force:

$$F_z = \frac{\dot{m}^2}{2\pi r_{\text{p}} t_{\text{sf}} \rho}. \tag{3.16}$$

In order to obtain a value of F_z, we estimated the pedestal radius r_p through the finite-element modelling by fitting resonance frequencies of the microtoroid mechanical modes observed in the experiment. The resonator used for the work presented in this chapter is highly undercut, with the pedestal radius estimated to be approximately 1.65 μm. Taking values for the superfluid film thickness and the absorbed heat power to be, as before, $t_{\text{sf}} = 2$ nm and $P_{\text{abs}} = 1\ \mu\text{W}$ respectively, we calculate $F_z = 5.8 \times 10^{-10}$ N.

As can be seen from Eqs. 3.16 and 3.3, the vertical force F_z should exhibit a quadratic dependence on the heat power P_{abs}. This has been experimentally observed for the lowest optical powers in seminal experiments in bulk superfluid helium [41, 42].

Even though in our experiments the magnitude of the vertical force F_z is comparable to that of the radial force F_{radial}, F_z is applied to a node of the displacement of the microtoroid mechanical mode of interest (see Sect. 3.3.1) and, therefore, does not efficiently couple to this mode. Thus, F_z does not play a significant role in our experiments, as also confirmed by the linear dependence of the measured superfluid photoconvective force on the laser power (Fig. 3.2b). However, we expect that in other experimental designs the vertical force F_z may be exploited beneficially.

3.3 Experimental Configuration and Results

3.3.1 Experimental Configuration

In order to investigate photoconvective forcing from superfluid flow, we employed the experimental configuration shown in Fig. 3.3a. A microtoroidal WGM optical cavity with major radius of $R_{\text{major}} = 37.5\ \mu\text{m}$ and minor radius of $R_{\text{minor}} = 3.5\ \mu\text{m}$ is enclosed within the sample chamber of a closed-cycle helium-3 cryostat. Fibre-laser light with wavelength of 1555.08 nm evanescently couples into a high-quality optical mode of the resonator via a tapered single-mode optical fibre. The linewidth of the microtoroid optical mode of interest is $\kappa/2\pi = 23.5$ MHz. Being a solid structure made of silica, the microtoroidal resonator exhibits intrinsic mechanical vibrations. Its motional modes span a frequency range from 1 MHz to 50 MHz. The thermal motion of these modes is encoded onto the optical field confined within the microtoroid as phase fluctuations which we resolve via balanced homodyne detection. The radial forces exerted onto the resonator both from the radiation pressure and the superfluid flow would be most efficient when acting on the radial breathing mode of the toroid due to an optimal overlap between the forces and the mode. In our structure this mode has a frequency of 40 MHz. However, we experimentally found that the forcing from the superfluid flow is ineffective above frequencies of approximately 2 MHz. We ascribe this behaviour to a possible breakdown of superfluidity which occurs as the superfluid critical velocity is reached [4]. The maximum frequency at which the superfluid can respond is limited by the critical velocity and the distance it needs to travel, which in our experiment is on the order of the microtoroid radius. Another possible reason for the bounded photothermal bandwidth could be the characteristic thermal response time of the resonator, which functions as a low-pass filter. As a consequence, for our experiments we chose the first-order flexural mode of the microtoroid with a frequency of $\Omega_{\text{m}}/2\pi = 1.35$ MHz and a mechanical dissipation rate of $\Gamma_{\text{m}}/2\pi = 530$ Hz at the base temperature of the cryostat (559 mK). This mode was chosen as it was observed to exhibit the strongest forcing from helium evaporation. The measured single-photon optomechanical coupling rate of this mode is $g_0/2\pi = 12.3$ Hz. Thermomechanical fluctuations of the flexural mode of interest at 3 K are shown in Fig. 3.3b. The inset of this figure illustrates the displacement profile of the mode obtained through the finite-element modelling.

In order to form a superfluid helium film on the microtoroid, the sample chamber is filled with helium-4 gas at low-pressure (19 mBar at 2.9 K). Upon cooling down towards the cryostat base temperature and crossing the superfluid phase transition temperature, the film naturally forms on the entire surface inside the sample chamber including the microtoroid. The low helium gas pressure is chosen deliberately in order to ensure a thickness of the superfluid film which would support third-sound modes with frequencies that do not overlap with the characteristic frequency of the microtoroid mechanical mode. At the chosen gas pressure the superfluid phase transition temperature is 850 mK, at which helium transitions from the gas directly into its superfluid state, forming an ultrathin (<5 nm) superfluid film.

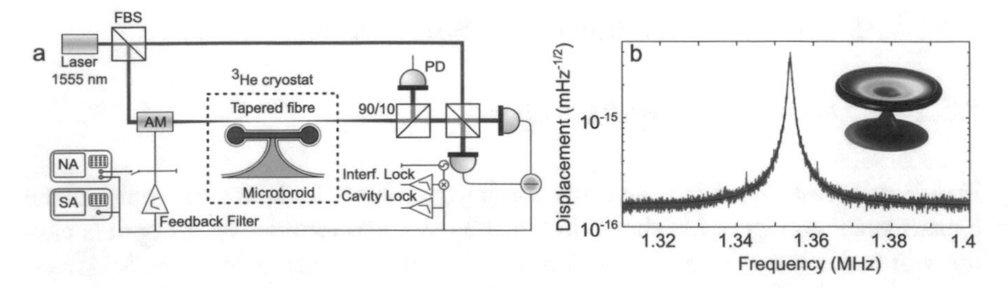

Fig. 3.3 **a** Experimental configuration employed to study the superfluid photoconvective forcing. A WGM optical cavity placed inside a sample chamber of a helium-3 cryostat forms a part of an all-fibre interferometer built for the balanced homodyne detection scheme. Electronic filters and amplification stages are implemented for the feedback cooling experiment described in Sect. 3.3.4. The interferometer and the cavity locking is utilized for stability of the interferometer phase and the laser frequency respectively. FBS—fibre beamsplitter, AM—amplitude modulator, SA—spectrum analyser, NA—network analyser, PD—photodetector. **b** Displacement profile due to thermomechanical motion of the microtoroid flexural mode of interest at 3 K. Inset: finite-element simulation of the mode's mechanical displacement profile. Figure is reproduced from Ref. [35] with modifications

3.3.2 Microtoroid Mode Thermometry

Knowing the temperature of the microtoroid is an important prerequisite for the work presented in this chapter. We start off by monitoring the displacement of the microtoroid flexural mode of interest while varying the cryostat temperature from 10 K down to 320 mK. Recording the homodyne photocurrent and performing its spectral analysis, we integrate the power spectral density of the mechanical mode and, thus, ascertain the mode's temperature. This yields an estimate of the thermal equilibrium final temperature of the mechanical mode. As can be seen from the linear fit in Fig. 3.4a, the microtoroid is well thermalized to its cryogenic environment in the temperature range from 10 K down to 560 mK. However, below 560 mK the microtoroid mechanical mode temperature plateaus and is no longer in thermal equilibrium with the cryostat (see Fig. 3.4a). We ascribe this temperature departure to the optical heat dissipated at the level of the microtoroid, resulting in a thermal gradient between the resonator and the cryostat cold plate.

We next investigate the effect of the heat generated due to optical absorption on the microtoroid temperature. To do so, we determine the temperature of the microtoroid mechanical mode as a function of the laser power. The temperature is obtained through integrating the power spectral density of the flexural mode and is observed to grow with rising laser power. Increasing optical heat results in thermal effects that degrade the system above certain threshold. In the case of our superfluid —based system this thermal degradation corresponds to boil-off of the superfluid film, as shown in Fig. 3.4b. First, while the laser power is increased over two orders of magnitude from 10 nW to 2.1 μW, the microtoroid mode temperature grows only modestly from 510 to 730 mK. However, when the laser power is further increased above 2.2 μW,

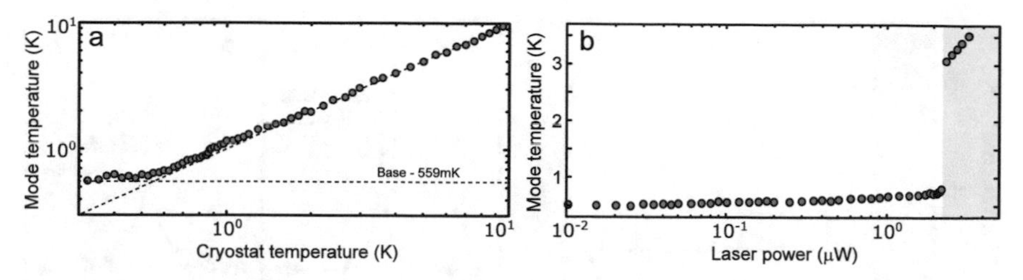

Fig. 3.4 **a** Temperature of the microtoroid flexural mechanical mode as a function of the cryostat temperature varied from 10 K down to 0.32 K. The plateau below 0.56 K is a microtoroid base temperature which it reaches with 100 nW of injected optical power. Below 0.56 K the microtoroid is no longer in thermal equilibrium with the cryostat due to the heat dissipated at the resonator. **b** Temperature of the microtoroid flexural mode as a function of the laser power varied from 10 nW up to 3.3 μW. Below 2.2 μW the mode temperature exhibits only a modest growth as the laser power is increased over two orders of magnitude from 10 nW to 2.1 μW. The abrupt temperature jump above 2.2 μW is caused by the superfluid film boil-off, resulting in the breakdown of the microtoroid thermalization to the cryostat. Figure is reproduced from Ref. [35] with minor modifications

the mechanical mode temperature exhibits an abrupt jump to about 3 K, which is indicated by the shaded area in Fig. 3.4b. This threshold is a consequence of the upper limit on the superfluid flow rate through the narrowest point of the pedestal which is dependent upon the optical heat power (see Sect. 3.2.2). The maximal flow rate through the junction between the pedestal and the microtoroid dictates the maximal optical thermal load which can be balanced out by the superfluid film. Plugging in the boil-off power of 2.2 μW into Eq. (3.15) for the flow velocity through the constriction point of the pedestal, yields ~30 m/s, that is in the vicinity of the superfluid critical velocity beyond which superfluidity breaks down and the superfluid flow ceases to be dissipationless. This causes a thermal run-away process, whereby the superfluid at the outer boundary of the resonator can no longer be replenished at the same rate as it evaporates and, hence, boils off entirely. As a result, thermalization of the microtoroid breaks and the resonator is no longer thermally anchored to the cryostat through the efficient superfluid link, but rather through the thermal conductance of its silicon pedestal, with its final temperature dominated by the optically generated heat.

3.3.3 Superfluid-Enhanced Optical Forces

The next experiment is designed to investigate microphotonic forces existing in our system. We coherently drive mechanical motion of the microtoroid flexural mode while varying the cryostat temperature and performing a network analysis (Fig. 3.5). A constant drive tone at the frequency of the flexural mode (1.35 MHz) is applied via amplitude modulation of the optical field coupled into the resonator. This drives the mechanical mode via resonant forces exerted on the microtoroid from both radi-

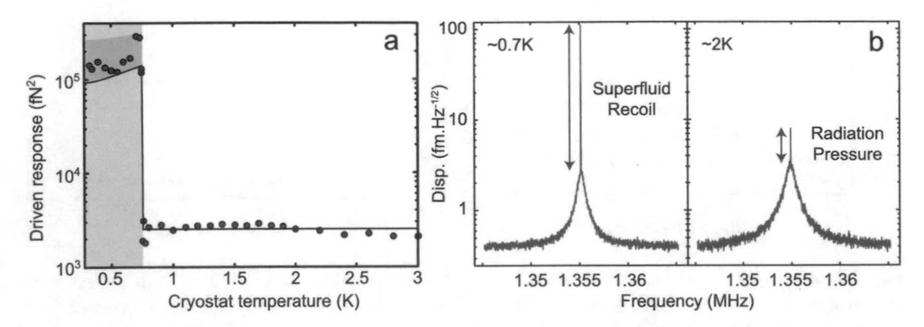

Fig. 3.5 **a** Measured response of the microtoroid flexural mode (red dots) to the coherent optical drive as the cryostat temperature is varied from 3 K down to ~320 mK. The experimental data exhibits a sharp increased response at the superfluid transition temperature. A theoretical fit to the data is shown in solid black line (theoretical fit has no free parameters). The grey-shaded region indicates the temperature range within which the microtoroid is covered with a superfluid film. The pink-shaded area shows a theoretically estimated band of the expected superfluid photoconvective force in the case if T_{evap} is up to 1 K higher than the mechanical mode temperature. **b** Comparison of the microtoroid flexural mode displacement spectra at 0.7 K (left) and 2 K (right) under a coherent optical drive applied via amplitude modulation of the injected light field. The driven response of the mechanical mode exhibits a large increase in the presence of the superfluid film. Figure is reproduced from Ref. [35] with minor modifications

ation pressure and, below the superfluid transition temperature, superfluid flow. Response of the flexural mode to this drive is imprinted onto the phase of the output light field which is measured via the homodyne detection scheme. Figure 3.5a illustrates how the mode's driven response varies with the cryostat temperature. Above the superfluid phase transition the optical force on the resonator comes solely from radiation pressure (right-hand side of Fig. 3.5b) and is basically independent of temperature, as indicated by the flat response of the mode in Fig. 3.5a. However, upon crossing the superfluid transition temperature and emergence of a superfluid film (grey-shaded area in Fig. 3.5a), the response of the mechanical mode to the coherent optical drive exhibits a sharp rise by 21 dB (Fig. 3.5a and b(left)). This manifests as an effective optical forcing of the flexural mode at the level of 540 fN. However, finite-element modelling reveals a very poor overlap between this mechanical mode and the radial recoil force from helium evaporation (see Sect. 3.2.1), with the calculated overlap being only 0.037%. Indeed, the flexural motion illustrated in the inset of Fig. 3.3 is predominantly out-of-plane, with only a minute radial motion component. The driving is most efficient when it is collinear with the motion. Thus, given the calculated overlap, we estimate that the 540 fN vertical drive produces a force of 1.46 nN applied radially. The boosted response of the microtoroid mechanical mode to the optical drive in the presence of superfluid well agrees with the theoretical estimations (Sect. 3.2.1) and corresponds to a superfluid photoconvective force that is eleven times stronger than its radiation pressure counterpart (Fig. 3.5a).

The combined forcing is well described by our theoretical model (see Sect. 3.2) which fits the experimental data without any free parameters (black curve in

Fig. 3.5a). As can bee seen from Fig. 3.5a, the experimental data exhibits a downward trend of applied photoconvective forcing below the superfluid transition temperature. This can be explained by a reduction in RMS velocities of the evaporated helium atoms (Eq. (3.9)), i.e. atoms at lower temperatures exert less recoil force on the resonator (Eq. (3.10)). This behaviour is well reproduced by our theoretical model, whereby we took the temperature of evaporated helium atoms in Eq. (3.10) to be equal to the measured microtoroid mechanical mode temperature T_m. However, as illustrated in Fig. 3.5a, the experimentally measured superfluid photoconvective forces are observed to systematically exceed their theoretically predicted values within maximum discrepancy of roughly 60%. We ascribe this deviation to a temperature imbalance between the evaporated helium atoms and the superfluid film. This is consistent with the past observations which demonstrated that, depending on the total heat applied to the liquid, helium atoms evaporated from a superfluid film tend to have a temperature up to 1 K higher than that of the superfluid film [40, 43]. In Fig. 3.5a this trend is illustrated by the theoretically estimated band (pink-shaded area) which shows the expected superfluid photoconvective force applied by helium atoms that are evaporated with temperatures T_{evap} in the range from the mode temperature T_{m} to $T_{\mathrm{m}} + 1$ K.

3.3.4 *Feedback Cooling of a Microtoroid Mechanical Mode*

In order to demonstrate an application that would benefit from the superfluid-enhanced optical forces, we damp motion of the microtoroid flexural mode through a feedback cooling protocol. Theoretical background of the feedback cooling has been elaborated in details in a wide range of publications [44–47]. Here, prior to describing the experimental results, we briefly provide the main postulates of the feedback cooling theory that are relevant to our experiments.

3.3.4.1 Theoretical Background of Feedback Cooling

The homodyne photocurrent, that in optomechanical systems is proportional to the optical field phase fluctuations induced by the motion of a mechanical element, can be quite generally represented as

$$\delta i(\omega) = G_{\mathrm{det}} \delta \tilde{x}(\omega) \tag{3.17}$$

$$= G_{\mathrm{det}} \left(\delta x(\omega) + \delta N(\omega) \right), \tag{3.18}$$

where $\delta N(\omega)$ is measurement noise and G_{det} is the detection gain provided by the detector, interferometer response and optomechanical interaction. The measurement noise in our system manifests as the optical shot noise that has a flat spectrum across the mechanics spectral region. In our experimental configuration the photocurrent is fed into an amplitude modulator located just before the microtoroidal optical cavity (see Fig. 3.3a). This leads to direct amplitude modulation of the intracavity light field which, in turn, results in an applied optically mediated feedback force proportional to the photocurrent, i.e. $F_{\mathrm{FB}}(\omega) \propto \delta i(\omega)$. Assembling the effects of detection, filtering and actuation into a combined *gain* term, we can write the feedback force as

$$F_{\mathrm{FB}}(\omega) = -g\chi^{-1}(\Omega_{\mathrm{m}})\delta\tilde{x}(\omega). \tag{3.19}$$

Here g represents the feedback gain and the term $\chi^{-1}(\Omega_{\mathrm{m}})$ ensures that the gain is unitless and eases factorization into the mechanical susceptibility in the consecutive steps. Plugging Eq. (3.19) into the quantum Langevin equation written in the Fourier domain yields for the oscillator position

$$\delta x(\omega) = \chi'(\omega)\left[F_{\mathrm{th}}(\omega) - g\chi^{-1}(\Omega_{\mathrm{m}})\delta N(\omega)\right] \tag{3.20}$$

where F_{th} is the thermal force and $\chi'^{-1}(\omega) = m_{\mathrm{eff}}^{-1}\left[\Omega_{\mathrm{m}}^2 - \omega^2 + i\Gamma'_{\mathrm{m}}\Omega_{\mathrm{m}}\right]^{-1}$ is the feedback-modified mechanical susceptibility with altered linewidth $\Gamma'_{\mathrm{m}}=\Gamma_{\mathrm{m}}\,(1+g)$. The integrated power spectral density obtained from the photocurrent as $S_{\tilde{x}\tilde{x}}(\omega) = \langle|\delta\tilde{x}(\omega)|^2\rangle$ can then give the mechanical mode temperature as

$$\tilde{T} = \int_{-\infty}^{\infty} \mathrm{d}\omega\, S_{\tilde{x}\tilde{x}}(\omega) \tag{3.21}$$

$$= \left(1 - \frac{g(g+2)}{\mathrm{SNR}}\right)\frac{1}{1+g}T_0. \tag{3.22}$$

Here T_0 is the initial temperature and SNR is the signal-to-noise ratio of the mechanical fluctuations peak to the noise of the detector signal that is used for feedback cooling. SNR is given by $\mathrm{SNR} = \frac{S_{\tilde{x}\tilde{x}}(\Omega_{\mathrm{m}})}{S_{\mathrm{NN}}(\Omega_{\mathrm{m}})}$. However, this is not the actual final temperature of the mechanical mode. Since $\tilde{T}$ is derived from the same photocurrent that is fed back into the microtoroid for feedback cooling, the in-loop photodetector shot noise, correlated with fluctuations of the microtoroid mechanical mode, is incorporated into $\tilde{T}$ [47]. Note although that these correlations occur only at high gain. The actual final mechanical mode temperature can be obtained by considering the signal from an independent out-of-loop photodetector which is uncorrelated with the signal from the in-loop detector prior to interaction with the microtoroid mechanical mode (see Ref. [47] for details). The actual mode temperature, independent of the feedback-induced correlations, is then given by

$$T = \frac{\mathrm{SNR} + g^2}{\mathrm{SNR} - g(g+2)}\tilde{T}. \tag{3.23}$$

3.3.4.2 Experimental Realization of Feedback Cooling

Experimentally the feedback cooling protocol in our system is realized by feeding the homodyne current into an amplitude modulator, that light passes through right before going into the optical cavity (Fig. 3.3a). In order to cool down the microtoroid mechanical mode, we have to ensure that the combined force from the superfluid flow and radiation pressure opposes the velocity of the mode. This requires the phase of the feedback loop to be adjusted appropriately, which is achieved by passing the homodyne current through a number of filter and amplification stages before it is fed into the amplitude modulator (Fig. 3.3a). This protocol allows us to cool the thermomechanical motion of the 1.35 MHz flexural microtoroid mode via cold damping [48]. The result is shown in Fig. 3.6a and b, which demonstrate that both the displacement amplitude of the flexural mode (Fig. 3.6b) and its temperature (Fig. 3.6a) decrease as the gain becomes stronger. This is in an excellent agreement with the theory presented above and in Ref. [46, 47]. With the initial temperature of $T_0 = 715$ mK in Eq. (3.22), $\tilde{T}$ accurately describes the behaviour of our system, as indicated by the solid line in Fig. 3.6a. However, $\tilde{T}$ gives the "in-loop" temperature estimate which is not the actual final mode temperature (see above in this § for the brief theoretical background). The actual mode temperature ("out-of-loop") is computed with Eq. (3.23) and shows that the microtoroid 1.35 MHz flexural mode is cooled from 715 to 137 mK (dashed line in Fig. 3.6a). The final mechanical occupancy of this mode is calculated to be $n = 2110 \pm 40$ phonons.

The final phonon occupancy of a mechanical mode, quite generally for the feedback cooling technique, is determined by both the optomechanical coupling rate and the ultimate strength of the feedback force [49]. The efficiency of the feedback cooling and the achievable temperature is limited by the level of motion at which the signal starts to merge with the detection noise (see Fig. 3.6b). With a

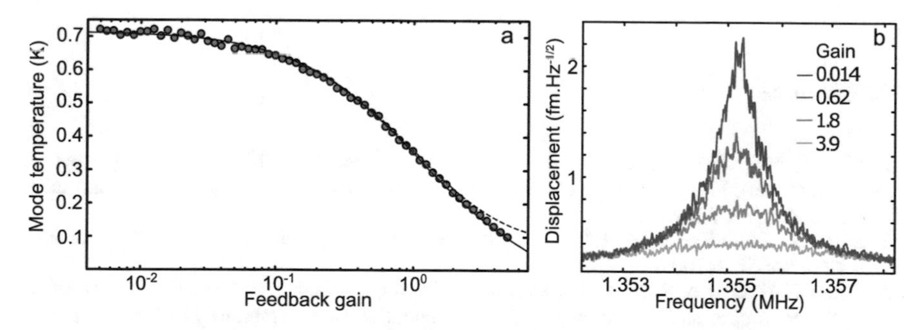

Fig. 3.6 **a** Temperature of the 1.35 MHz microtoroid flexural mode as a function of the feedback gain. The mode is cooled from 715 to 137 mK via superfluid-mediated cold damping within a feedback loop. The probe power is kept at 1.9 μW while the feedback gain is swept over three orders of magnitude. The black solid line shows the mode temperature estimate from in-loop measurements. The dashed line represents the out-of-loop mode temperature computed with Eq. (3.23). **b** The microtoroid flexural mode displacement spectrum with varying feedback gain. Figure is reproduced from Ref. [35] with minor modifications

larger optomechanical coupling rate, this occurs at a lower effective temperature and smaller displacement amplitude, enabling more effective cooling. In our system the constraint is imposed by not-so-high optomechanical coupling rate of the microtoroid flexural mode which ultimately limits the final phonon occupancy. For the parameters realized in our system—choice of the mechanical mode, cryostat temperature, and the maximum feedback force magnitude of 1.46 nN—we estimate that the superfluid-mediated feedback force would become a constraint for a mechanical occupancy only below $n \approx 0.015$. Thus, we expect that cooling of mechanical modes to their quantum ground states should be possible with superfluid-enhanced photoconvective forces, given that the microtoroid parameters are improved to the level of those demonstrated in Refs. [24, 50]. Moreover, combining the superfluid-mediated feedback forces with back-action evading [18–20] or nonlinear measurements [21, 22], generation of non-classical mechanical states should also be achievable.

3.4 Conclusion

We demonstrated a novel approach to microphotonic forcing of mechanical elements which is based on photoconvective flow of superfluid helium and recoil. Our theoretical calculations and experiments showed that, in cryogenic conditions, forces significantly stronger than radiation pressure are available using superfluid photoconvective effects. The frictionless flow of superfluid helium allows fast actuation of mechanical elements to be realized in addition to their strong thermal anchoring, enabling relatively high optical powers to be used in cryogenic environments. Furthermore, in appropriately designed optomechanical systems, superfluid flow forces could be used to exert e.g. torques, and to stretch or compress microphotonic elements, providing a versatile tool to actuate cryogenic microphotonic circuits.

References

1. Chan J, Alegre TPM, Safavi-Naeini AH, Hill JT, Krause A, Gröblacher S, Aspelmeyer M, Painter O (2011) Laser cooling of a nanomechanical oscillator into its quantum ground state. Nature 478(7367):89–92
2. Metzger CH, Karrai K (2004) Cavity cooling of a microlever. Nature (London) 432(7020):1002–1005
3. Usami K, Naesby A, Bagci T, Nielsen BM, Liu J, Stobbe S, Lodahl P, Polzik ES (2012) Optical cavity cooling of mechanical modes of a semiconductor nanomembrane. Nat Phys 8(2):168–172
4. Tilley DR, Tilley J (1990) Superfluidity and superconductivity. CRC Press
5. Li M, Pernice WHP, Xiong C, Baehr-Jones T, Hochberg M, Tang HX (2008) Harnessing optical forces in integrated photonic circuits. Nature 456(7221):480
6. Roels J, De Vlaminck I, Lagae L, Maes B, Van Thourhout D, Baets R (2009) Tunable optical forces between nanophotonic waveguides. Nat Nanotechnol 4(8):510–513
7. Chu S, Bjorkholm JE, Ashkin A, Cable A (1986) Experimental observation of optically trapped atoms. Phys Rev Lett 57(3):314–317

8. Ashkin A, Dziedzic JM (1987) Optical trapping and manipulation of viruses and bacteria. Science 235(4795):1517–1520
9. MacDonald MP, Spalding GC, Dholakia K (2003) Microfluidic sorting in an optical lattice. Nature 426(6965):421–424
10. Burg TP, Godin M, Knudsen SM, Shen W, Carlson G, Foster JS, Babcock K, Manalis SR (2007) Weighing of biomolecules, single cells and single nanoparticles in fluid. Nature 446(7139):1066–1069
11. Taylor MA, Janousek J, Daria V, Knittel J, Hage B, Bachor HA, Bowen WP (2013) Biological measurement beyond the quantum limit. Nat Photonics 7(3):229–233
12. Metcalfe M (2014) Applications of cavity optomechanics. Appl Phys Rev 1(3):031105
13. Mamin HJ, Rugar D (2001) Sub-attonewton force detection at millikelvin temperatures. Appl Phys Lett 79(20):3358–3360
14. Forstner S, Prams S, Knittel J, van Ooijen ED, Swaim JD, Harris GI, Szorkovszky A, Bowen WP, Rubinsztein-Dunlop H (2012) Cavity optomechanical magnetometer. Phys Rev Lett 108(12):120801
15. Chaste J, Eichler A, Moser J, Ceballos G, Rurali R, Bachtold A (2012) A nanomechanical mass sensor with yoctogram resolution. Nat Nanotechnol 7(5):301–304
16. Beugnon J, Tuchendler C, Marion H, Gaetan A, Miroshnychenko Y, Sortais YRP, Lance AM, Jones MPA, Messin G, Browaeys A, Grangier P (2007) Two-dimensional transport and transfer of a single atomic qubit in optical tweezers. Nat Phys 3(10):696–699
17. Wilson DJ, Sudhir V, Piro N, Schilling R, Ghadimi A, Kippenberg TJ (2015) Measurement-based control of a mechanical oscillator at its thermal decoherence rate. Nature (London), advance online publication
18. Braginsky VB, Vorontsov YI, Thorne KS (1980) Quantum nondemolition measurements. Science 209(4456):547–557
19. Wollman EE, Lei CU, Weinstein AJ, Suh J, Kronwald A, Marquardt F, Clerk AA, Schwab KC (2015) Quantum squeezing of motion in a mechanical resonator. Science 349(6251):952–955
20. Szorkovszky A, Doherty AC, Harris GI, Bowen WP (2011) Mechanical squeezing via parametric amplification and weak measurement. Phys Rev Lett 107(21):213603
21. Thompson JD, Zwickl BM, Jayich AM, Marquardt F, Girvin SM, Harris JGE (2008) Strong dispersive coupling of a high-finesse cavity to a micromechanical membrane. Nature 452(7183):72–75
22. Brawley GA, Vanner MR, Larsen PE, Schmid S, Boisen A, Bowen WP (2016) Nonlinear optomechanical measurement of mechanical motion. Nat Commun 7:10988
23. Brooks DWC, Botter T, Schreppler S, Purdy TP, Brahms N, Stamper-Kurn DM (2012) Non-classical light generated by quantum-noise-driven cavity optomechanics. Nature 488(7412):476–480
24. Verhagen E, Deléglise S, Weis S, Schliesser A, Kippenberg TJ (2012) Quantum-coherent coupling of a mechanical oscillator to an optical cavity mode. Nature 482(7383):63–67
25. Rosenberg J, Lin Q, Painter O (2009) Static and dynamic wavelength routing via the gradient optical force. Nat Photonics 3(8):478–483
26. Restrepo J, Gabelli J, Ciuti C, Favero I (2011) Classical and quantum theory of photothermal cavity cooling of a mechanical oscillator. C R Phys 12(9–10):860–870
27. Meenehan SM, Cohen JD, MacCabe GS, Marsili F, Shaw MD, Painter O (2015) Pulsed excitation dynamics of an optomechanical crystal resonator near its quantum ground state of motion. Phys Rev X 5(4):041002
28. Wu M, Hryciw AC, Healey C, Lake DP, Jayakumar H, Freeman MR, Davis JP, Barclay PE (2014) Dissipative and dispersive optomechanics in a nanocavity torque sensor. Phys Rev X 4(2):021052
29. Rudnick I, Kojima H, Veith W, Kagiwada RS (1969) Observation of superfluid-helium persistent current by doppler-shifted splitting of fourth-sound resonance. Phys Rev Lett 23(21):1220–1223
30. Rokhsari H, Vahala KJ (2004) Ultralow loss, high Q, four port resonant couplers for quantum optics and photonics. Phys Rev Lett 92(25):253905

31. Bagheri M, Poot M, Li M, Pernice WPH, Tang HX (2011) Dynamic manipulation of nanomechanical resonators in the high-amplitude regime and non-volatile mechanical memory operation. Nat Nanotechnol 6(11):726–732
32. Baker C, Hease W, Nguyen D-T, Andronico A, Ducci S, Leo G, Favero I (2014) Photoelastic coupling in gallium arsenide optomechanical disk resonators. Opt Express 22(12):14072–14086
33. Chase CE (1962) Thermal conduction in liquid helium II. I. Temperature dependence. Phys Rev 127(2):361
34. Zeller RC, Pohl RO (1971) Thermal conductivity and specific heat of noncrystalline solids. Phys Rev B 4(6):2029
35. McAuslan DL, Harris GI, Baker C, Sachkou Y, He X, Sheridan E, Bowen WP (2016) Microphotonic forces from superfluid flow. Phys Rev X 6(2):021012
36. Atkins KR (1959) Third and fourth sound in liquid Helium II. Phys Rev 113(4):962–965
37. Donnelly RJ, Barenghi CF (1998) The observed properties of liquid helium at the saturated vapor pressure. J Phys Chem Ref Data 27(6):1217–1274
38. Scholtz JH, McLean EO, Rudnick I (1974) Third sound and the healing length of He II in films as thin as 2.1 atomic layers. Phys Rev Lett 32(4):147–151
39. Jeans J (1982) An introduction to the kinetic theory of gases. Cambridge University Press, Cambridge Cambridgeshire
40. Hyman DS, Scully MO, Widom A (1969) Evaporation from superfluid helium. Phys Rev 186(1):231–238
41. Hall HE (1954) The inertia of heat flow in liquid helium II. Proc Phys Soc A 67(6):485
42. Kapitza PL (1964) Collected papers of PL Kapitza. Pergamon Press
43. Andres K, Dynes RC, Narayanamurti V (1973) Velocity spectrum of atoms evaporating from a liquid He surface at low temperatures. Phys Rev A 8(5):2501–2506
44. Pinard M, Cohadon PF, Briant T, Heidmann A (2000) Full mechanical characterization of a cold damped mirror. Phys Rev A 63(1):013808
45. Poggio M, Degen CL, Mamin HJ, Rugar D (2007) Feedback cooling of a cantilever's fundamental mode below 5 mK. Phys Rev Lett 99(1):017201
46. Lee K, McRae T, Harris G, Knittel J, Bowen W (2010) Cooling and control of a cavity optoelectromechanical system. Phys Rev Lett 104(12):123604
47. Harris GI (2014) Cavity optomechanics with feedback and fluids. PhD thesis, The University of Queensland
48. Cohadon PF, Heidmann A, Pinard M (1999) Cooling of a mirror by radiation pressure. Phys Rev Lett 83(16):3174–3177
49. Bowen WP, Milburn GJ (2015) Quantum optomechanics. CRC Press
50. Anetsberger G, Rivière R, Schliesser A, Arcizet O, Kippenberg TJ (2008) Ultralow-dissipation optomechanical resonators on a chip. Nat Photonics 2(10):627–633

Chapter 4
Theoretical Investigation of Vortex-Sound Interactions in Two-Dimensional Superfluids

Vortices play an essential role in two-dimensional superfluids since they are responsible for the onset of superfluidity in these systems, as described by the theory of Berezinskii-Kosterlitz-Thouless phase transition [1–4], and determine the dynamics of superfluids, giving rise to dissipation [5, 6], quantum turbulence [7], and generation of exotic vortex configurations [8]. Interaction between vortices and sound is of broad significance in Bose-Einstein condensates of dilute gases and superfluid helium [9–13]. However, quantifying and modelling the vortex flow field and its interaction with sound are sophisticated hydrodynamic problems, with analytic solutions available only in special cases [14]. In this work we develop methods to compute both the vortex and sound flow fields within an arbitrary two-dimensional domain. Furthermore, we theoretically investigate the dispersive vortex-sound interaction in two-dimensional superfluids and develop a model which quantifies this interaction for any vortex distribution on any two-dimensional bounded domain, not necessarily simply-connected. To achieve this, we utilize analogies between superfluid and vortex dynamics and respectively fluid dynamics of an ideal gas and electrostatics. Using this technique, we propose an experiment leading to an unambiguous detection of single circulation quanta in two-dimensional superfluid helium.[1]

4.1 Introduction

Superfluidity in two-dimensional matter has been subject to extensive theoretical and experimental research efforts since the 70's of the past century, culminating with the 2016 Physics Nobel Prize, awarded for understanding the nature of superfluidity

[1] This chapter is based on the work published by IOP Publishing: S. Forstner, **Y. Sachkou**, M. Woolley, G. I. Harris, X. He, W. P. Bowen, C. G. Baker, Modelling of vorticity, sound and their interaction in two-dimensional superfluids, *New Journal of Physics* 21, 053029, 2019.

Y. Sachkou, *Probing Two-Dimensional Quantum Fluids with Cavity Optomechanics*, Springer Theses, https://doi.org/10.1007/978-3-030-52766-2_4

in two dimensions [1, 2, 4]. First systematically studied in thin films of superfluid helium [15–18], superfluid phase transitions have been discovered in a great variety of other two-dimensional physical systems, such as Bose-Einstein condensates [19, 20] (BEC), topological superconductors [21], exciton-polariton condensates [22, 23] etc. Quantized vortices—elementary excitations existing in various quantum fluids and superconductors—play an essential role in all these condensed matter systems, as they trigger the onset of phase transitions in two-dimensional matter. Thus, binding of quantized vortices into pairs enables the emergence of long-range phase order and transition of liquid helium into its superfluid phase in thin helium films. Quantifying and modelling the interaction of vortices with phonons and rotons—other forms of elementary excitations in quantum fluids which manifest as sound—is of great importance for understanding such phenomena as quantum turbulence [10, 24–26] and dissipation induced by vortex-sound interactions. It is also crucial for the ability to experimentally control vortices with sound and for tracking vortex dynamics. Therefore, the capability to determine the flow field generated by an arbitrary vortex configuration on an arbitrary and perhaps multiply-connected geometry is of broad importance, as it enables to quantify the strength of the sound-vortex coupling.

The relevant sound eigenmodes in BECs are density- and temperature-waves, called respectively *first-* and *second sound*. In the case of two-dimensional films of superfluid helium, modes of both first- and second sound are suppressed due to the incompressibility of the fluid and the clamping of the normal fluid component. Surface waves (*third-sound*—see Sect. 1.10) is the primary form of sound modes in thin films of superfluid helium [27]. Recently, both propagation of temperature-waves in a two-dimensional Bose-Einstein condensate [28] and real-time measurement and control of third-sound modes in superfluid helium thin films [29] have been demonstrated. However, up till now there have been no available numerical tools for modelling sound-vortex interactions within an arbitrary domain and for an arbitrary distribution of vortices.

In this work we develop a perturbative analytical model for two-dimensional superfluids, allowing to compute the rate of vortex-sound interactions for any arbitrary distribution of vortices within a circular domain. We provide an intuitive insight to how vortex flow field induces degeneracy lifting and frequency splitting of sound modes. Moreover, we go further and develop a finite-element method to model vortex flow fields and the interaction of sound waves with vortices in a two-dimensional superfluid by utilising analogies with other areas of physics. We map vortex dynamics onto electrostatics and superfluid hydrodynamics onto fluid dynamics of an ideal gas and leverage well-established finite-element modelling (FEM) tools available for these fields. We show how the interaction of an arbitrarily distributed vortex ensemble with sound on any arbitrary two-dimensional geometry, not necessarily simply-connected, can be modelled using these tools. We discuss the interaction of sound modes with vortices within an experimentally relevant disk-shaped domain. This case is relevant for a number of experiments with thin films of superfluid helium [12, 29–31] (also see Chaps. 2, 3 and 5 of this thesis) and two-dimensional Bose-Einstein-condensates [28, 32, 33]. Furthermore, we compare the results of our analytical model for the sound-vortex interactions with FEM results and show that the analytical model provides a good approximation to the FEM simulation within circular geometries.

In the last part of this chapter we focus our analysis on the prospect of detecting a single quantum of circulation in two-dimensional superfluid helium. While vortices in Bose-Einstein condensates can be experimentally visualized by optical snapshots [34], and vortices in exciton-polariton condensates can be seen by optical interferometry, no such direct observation technique is available for thin superfluid helium films. The reason behind this is that the vortex core in superfluid helium is an Ångström-size perturbation to an ultra-thin film of transparent liquid, whose flow does not interact dissipatively with the environment. Measurements of vortex dynamics in superfluid helium are important because, unlike in most BECs of dilute gases and exciton-polariton condensates where atom-atom interactions are weak, helium atoms exhibit strong atom-atom interactions and dynamics of superfluid helium cannot be well approximated by the Gross-Pitaevskii-ansatz, and are not fully understood [35].

In seminal works by Ellis et al. [11, 12, 36], the interaction of a large number of vortices with third-sound modes on a centimeter-scale circular resonator was observed experimentally. Moreover, this interaction was modelled assuming centered vortex distributions [12, 36, 37]. However, more precise tracking and observation of vortex dynamics was hindered by both the experimental capacity to resolve the effects of smaller vortex numbers and by the ability to quantify the interaction of sound modes with individual vortices arbitrarily located on a circular resonator.

Here, we propose an experiment where discrete steps in sound modes frequency splittings due to an increase/decrease of a number of circulation quanta could be observed for the case of superfluid helium thin films. We propose a structure where pinning of quanta of circulation around an engineered topological defect leads to experimentally observable quantized steps. Utilising our finite-element model, we discuss how, in the proposed geometry, the sound-vortex interaction can be maximized, so that the quantized steps could be clearly resolved. This would enable the first detection of a single quantum of circulation in two-dimensional superfluid helium.

4.2 Analytical Investigation of Sound-Vortex Interactions

The interaction between sound waves and quantized vortices in two-dimensional superfluids can be studied by considering the kinetic energy landscape of the system with and without vortices. We can understand this interaction through the change in sound wave kinetic energy induced by an addition or subtraction of a vortex. The irrotational flow fields of sound modes are orthogonal to vortex flow fields, which are fully defined by the fluid rotation around the vortex core, Fig. 4.1a (see Eq. 4.29). Hence, this yields a zero overlap between vortex and sound velocity fields:

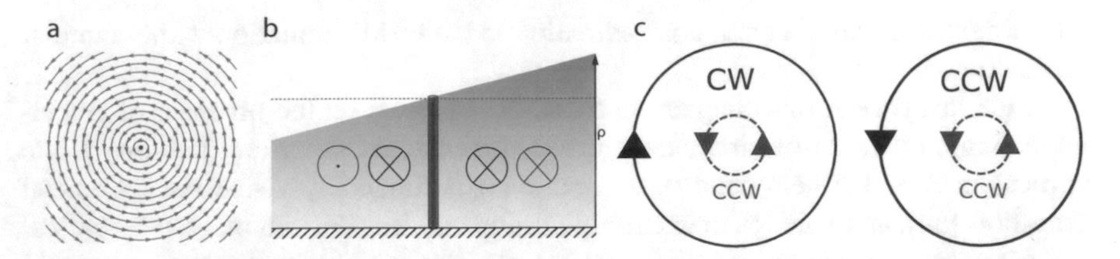

Fig. 4.1 **a** Streamlines of a point vortex (red dot) in a two-dimensional domain. The circulation of the vortex flow field is equal to the circulation quantum κ in any loop encompassing the core and is zero elsewhere. **b** Illustration of the coupling between vortex- (red) and third-sound mode (black) flow fields. The red spiral in the middle represents the vortex. Destructive interference on the left from the vortex and constructive interference on its right, in the presence of a mode-induced superfluid film height gradient, causes an interaction between the vortex and third sound (see text for more details). The vertical axis can refer to density in a Bose-Einstein condensate or the film height in a superfluid helium thin film. **c** Perturbation to clockwise (CW, black) and counter-clockwise (CCW, black) third-sound modes caused by the presence of a vortex with a given direction of circulation (counter-clockwise, red). The sound mode co-rotating with the vortex flow field (CCW mode) gains energy, whereas CW sound mode loses energy

$$\int_A \vec{v}_v \cdot \vec{v}_s \, dA = 0, \tag{4.1}$$

where A is the area of the domain where both vortices and sound are confined to, $\vec{v}_v$ is the vortex velocity field and $\vec{v}_s$ is the two-dimensional sound velocity distribution. Equation (4.1) indicates that there is no coupling between the two flow fields. However, the interaction occurs due to the change in film height (for superfluid helium) or density (for BEC) associated with the sound wave. Locally, sound and vortex flows add up to change the kinetic energy of the fluid. This is illustrated in Fig. 4.1b, where a density/height increase on one side of the vortex causes a kinetic energy imbalance. Namely, the increased kinetic energy due to sound and vortex velocities addition on the right side of the vortex is not fully compensated by the reduction in kinetic energy due to velocity subtraction on the left, resulting in a net increase in kinetic energy due to the sound-vortex interaction.

Here we analytically investigate the sound-vortex interaction and derive the frequency splitting experienced by a third-sound mode in the presence of a single quantized vortex. In order to compute the splitting, we consider the flow field of a single vortex as a perturbation to the third-sound mode confined within a circular resonator. We would also like to point out that even though in the following, for simplicity, we develop an analytical framework for thin films of superfluid helium (i.e. third-sound waves), the analysis can also be applied to Bose-Einstein condensates with a straightforward replacement of variables (see Appendix A.1).

Third-sound modes are solutions of the wave equation and on a circular resonator are characterized by Bessel modes of the first kind (see Sect. 1.10). Modes are fully quantified by their azimuthal ($m \geq 0$) and radial ($n \geq 1$) orders, which define the node counts along the circumference and radius respectively. Modes with $m \neq 0$

can be represented in a basis of two travelling modes rotating in opposite directions: clockwise (CW) and counter-clockwise (CCW). The complex surface displacement amplitudes η for travelling CW and CCW third-sound waves (or alternatively the sound-induced density fluctuations for a Bose-Einstein condensate) in a circular domain are given by [38]

$$\eta\left(r,\theta,t\right) = \eta_0\, J_m\left(\xi_{m,n}\frac{r}{R}\right) e^{i\left(m\theta\pm\Omega\, t\right)}, \tag{4.2}$$

where η_0 is the wave amplitude, J_m is the Bessel function of the first kind of order m, and m and n are respectively the azimuthal and radial mode orders; $\xi_{m,n}$ is a frequency parameter depending on the mode order and the boundary conditions [38]. Here '+' and '−' signs respectively correspond to the CW and CCW travelling waves. In the absence of any external perturbations and in a perfect circular domain with no geometrical defects, CW and CCW modes are degenerate. The presence of a vortex, placed in the same circular domain, manifests as a perturbation to the CW and CCW degenerate modes. Therefore, the presence of a vortex lifts the degeneracy (Fig. 4.1c). Without loss of generality, here we assume a vortex with a counter-clockwise (CCW) direction of circulation. Thus, the CW third-sound mode, travelling in the opposite direction relative to the vortex flow field, loses energy and is frequency down-shifted (Fig. 4.1c(left)). Whereas the CCW sound mode, co-rotating with the vortex flow, gains kinetic energy and is frequency up-shifted (Fig. 4.1c(right)). This frequency splitting Δf between CW and CCW third-sound modes is experimentally resolved if it is greater than the decay rate Γ of the sound mode (see Fig. 4.4b).

In order to compute the vortex-induced third-sound mode splitting, we first have to quantify both vortex and third-sound flow fields, which are elaborated below.

4.2.1 Point Vortex Flow Field $\vec{v}_v$

To compute a point vortex flow field $\vec{v}_v$, we utilise the potential flow theory. In a fluid, trajectories of the constituent particles in a steady flow are indicated by streamlines, which are tangential to the fluid velocity vector everywhere within the fluid. This means that the fluid flow never intersects a streamline. For irrotational, i.e. potential, and incompressible two-dimensional flows, the streamfunction is a scalar quantity which is constant along the streamlines and varies with displacements away from the streamlines. The flow velocity components are expressed as partial spatial derivatives of the streamfunction, which makes the latter a very useful quantity for computing the vortex flow field. The streamfunction Ψ of a point vortex with circulation κ in a two-dimensional plane is given by $\Psi = -\frac{\kappa}{2\pi}\ln\left(r\right)$. Our goal is to compute a point vortex flow field within a circular domain, which is a well-known problem [39]. The streamfunction Ψ for such configuration in Cartesian coordinates is given by:

$$\Psi = -\frac{\kappa}{2\pi}\left(\ln\left(\sqrt{(x-X_1)^2+y^2}\right) - \ln\left(\sqrt{(x-X_2)^2+y^2}\right)\right). \tag{4.3}$$

The first term represents the streamfunction of a point vortex inside a circular domain, radially offset by X_1 from the disk origin along x axis. The second term stands for the streamfunction of the opposite circulation image-vortex. The inclusion of the image-vortex streamfunction is required to enforce the no-flow across the domain boundary [39]. $X_2 = \frac{R^2}{X_1}$ is the radial coordinate of the image vortex.

The streamfunction Ψ now allows us to compute the vortex velocity components:

$$v_{vx} = \frac{\partial \Psi}{\partial y} \quad \text{and} \quad v_{vy} = -\frac{\partial \Psi}{\partial x}. \tag{4.4}$$

Using Eq. (4.4), in Fig. 4.2a we plot the flow streamlines of a point vortex offset by $R/2$ from the disk origin inside a circular resonator of radius R.

4.2.2 Third-Sound Flow Field $\vec{v}_3$

The third-sound surface displacement amplitude η is given by the Eq. (4.2). And the modes velocity profiles $\vec{v}_3$ are obtained as

$$\vec{v}_3 = \pm \frac{i\, c_3^2}{\Omega\, h_0} \vec{\nabla}\eta, \tag{4.5}$$

where, as for the displacement η, signs '+' and '−' respectively stand for CW and CCW travelling waves. Here, c_3 is the phase velocity of third-sound, Ω is its mechanical frequency, and h_0 the mean thickness of the superfluid film. We would like to note that while the displacement of a solid circular membrane would also be given by Eq. (4.2), its velocity would be different to Eq. (4.5), leading to significantly different effective mass scalings compared to third sound [38] (see Sect. 2.2.2). As an example, the surface displacement profile Re (η) and instantaneous velocity field Re $(\vec{v}_3)$ of a CW $(m = 1,\ n = 2)$ third-sound mode with free boundary conditions are plotted in Fig. 4.2b.

While the flow of the third-sound mode illustrated in Fig. 4.2b is irrotational and, hence, not associated with any in-plane circulation ($\oint \vec{v} \cdot \vec{dl} = 0$ for any closed loop inside the superfluid), it is associated with a net mass flow (for the CW travelling mode, there is more fluid moving clockwise under the wave crest (red) than counter-clockwise under the trough (blue), and similarly there is net CCW fluid motion for the CCW mode) (Fig. 4.2). It is this net mass flow which couples to the vortex field, resulting in a higher kinetic energy for the sound mode travelling in the same direction as the vortex flow.

Having quantified the vortex and third-sound flow fields, we are now set to compute the vortex-induced third-sound mode frequency splitting.

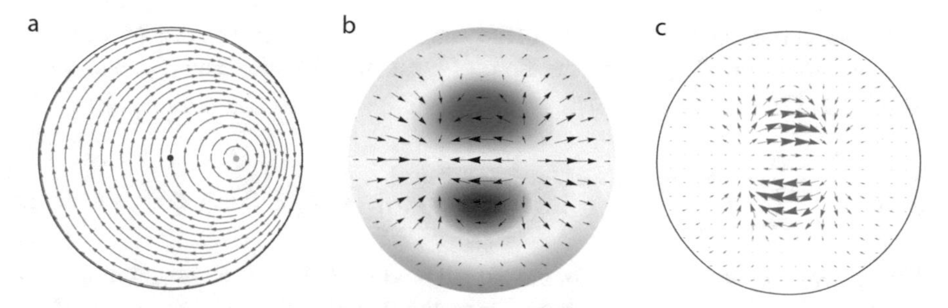

Fig. 4.2 **a** Vortex flow field $\vec{v}_v(\vec{r})$ streamlines for a clockwise (CW) vortex (green dot) offset from the disk origin (red dot) by $R/2$ in a circular domain of radius R. **b** Black arrows indicate the instantaneous velocity field Re $(\vec{v}_3)$ of a clockwise-rotating third-sound Bessel mode ($m = 1$; $n = 2$) with free boundary condition. The surface color-code represents the associated surface displacement profile Re (η) of this third-sound mode (color code: red—positive, blue—negative). Clockwise rotation of the mode can be seen by noticing that fluid starting to accumulate ahead of the red crest, where $\vec{\nabla} \cdot \vec{v}_3 < 0$, and emptying ahead of the blue trough, where $\vec{\nabla} \cdot \vec{v}_3 > 0$. The velocity field is positive under the crests, negative under the troughs, and irrotational, i.e. $\oint \vec{v}_3 \cdot \vec{dl} = 0$ for all contours inside the superfluid. **c** Vector field of $\vec{v}_3(\vec{r}) \times \eta(\vec{r})$. Combining velocity fields of a vortex (**a**) and a third-sound mode (**b**), one can see that $\iint \vec{v}_3 \cdot \vec{v}_v = 0$ (this can also be seen from a symmetry argument). However, multiplication by the surface displacement profile Re (η) results in a non-zero energy shift of the CW/CCW third-sound waves (see Eq. (4.9))

4.2.3 Analytical Derivation of the Vortex-Induced Sound-Mode Splitting

Here we analytically investigate how the vortex flow field modifies the kinetic energy landscape of the sound mode. We note that in this work we describe the quasi-static regime where the motion of vortices during a sound oscillation period is negligible. To verify the validity of this approximation, we estimate the orbit period T for a single vortex within a circular resonator of radius R, offset from the disk center by a distance x. This period T is given by $T = \frac{4\pi^2}{\kappa}\left(R^2 - x^2\right)$. For instance, for resonator with $R \simeq 10^{-5}$ m, this corresponds to typical Hz vortex orbit frequencies compared to typical 10^5 Hz third-sound frequencies [29]. The quasi-static regime applies in the limit of pinned vortices [5, 12, 37] or low vortex densities, where the velocity of the flow field generated by neighbouring and image vortices is significantly smaller than the speed of sound. And for the case of BECs the quasi-static approximation is valid if the vortex orbit period is substantially longer than the period of collective atomic oscillations.

Fluid flow kinetic energy within a cylindrically symmetric domain is given by

$$E = \frac{1}{2}\int \rho\, \vec{v}^{\,2}(\vec{r})\, \mathrm{d}^3(\vec{r})\,, \tag{4.6}$$

where ρ is the fluid density and $\vec{v}$ its velocity.

We use this expression to compute the kinetic energy difference $\Delta E\,(t)$ between a sound wave travelling along the flow of a quantized vortex and a sound wave travelling against the vortex flow:

$$\Delta E\,(t) = \frac{1}{2}\rho \int_{\theta=0}^{2\pi}\int_{r=0}^{R}\int_{z=0}^{h_0+\eta(r,\,\theta,\,t)} \left(||\vec{v}_3\,(\vec{r},\,t) + \vec{v}_v\,(\vec{r})||^2 - ||\vec{v}_3\,(\vec{r},\,t) - \vec{v}_v\,(\vec{r})||^2\right) r\,\mathrm{d}r\,\mathrm{d}\theta\,\mathrm{d}z. \tag{4.7}$$

This general expression works for any sound mode and any vortex position in a circular domain. We then make a reasonable assumption that $\vec{v}_3$ and $\vec{v}_v$ are independent of z, as the inviscid nature of the superfluid precludes any in-plane vorticity and does not require cancellation of the horizontal velocity at $z = 0$ (*no-slip* boundary). Given that in the absence of the vortex flow field the basis sound modes (Eq. (4.2)) are degenerate and, thus, magnitudes of their velocity fields are equal, Eq. (4.7) becomes:

$$\Delta E\,(t) = 2\rho \int_{\theta=0}^{2\pi}\int_{r=0}^{R} \vec{v}_3\,(r,\theta,\,t)\cdot\vec{v}_v\,(r,\theta)\,(h_0 + \eta\,(r,\theta,t))\,r\,\mathrm{d}r\,\mathrm{d}\theta. \tag{4.8}$$

We now notice that v_{3x} and v_{vx} as well v_{3y} and v_{vy} are functions of θ of different parity (see Fig. 4.2), i.e. v_{3x} is an even function of angle θ while v_{vx} is odd, and v_{3y} is odd while v_{vy} is even. This implies that $\iint \vec{v}_3\cdot\vec{v}_v = 0$, and Eq. (4.8) takes the following shape:

$$\Delta E\,(t) = 2\rho \int_{\theta=0}^{2\pi}\int_{r=0}^{R} \vec{v}_3\,(r,\theta,\,t)\cdot\vec{v}_v\,(r,\theta)\,\eta\,(r,\theta,t)\,r\,\mathrm{d}r\,\mathrm{d}\theta. \tag{4.9}$$

This can be understood as a form of surface-averaged planar Doppler shift, weighted by the displacement amplitude η of the sound mode.

Given that the sound mode velocity field $\vec{v}_3$ is a function of time, we next consider the time-averaged energy difference $\langle\Delta E\rangle$, averaged over a sound oscillation period T:

$$\langle\Delta E\rangle = \frac{1}{T}\int_0^T \Delta E\,(t)\,\mathrm{d}t = 2\,\rho \int_r\int_\theta r\,\mathrm{d}r\,\mathrm{d}\theta \left(v_{v\,r}\,\frac{1}{T}\int_0^T v_{3\,r}\,\eta\,\mathrm{d}t\right) + \left(v_{v\,\theta}\,\frac{1}{T}\int_0^T v_{3\,\theta}\,\eta\,\mathrm{d}t\right), \tag{4.10}$$

where $\vec{v}_3$ and $\vec{v}_v$ are expressed in terms of their radial and angular components, i.e. $v_{3\,r}$ and $v_{3\,\theta}$, and $v_{v\,r}$ and $v_{v\,\theta}$ respectively. Third-sound mode radial and angular velocity components in cylindrical coordinates are given by

$$v_{3\,r} = i\,\eta_0\,\frac{c_3^2\,R}{\xi_{m,n}\,h_0}\,\frac{\partial}{\partial r} J_m\left(\xi_{m,n}\frac{r}{R}\right)\,e^{i(m\theta\pm\Omega t)}, \tag{4.11}$$

$$v_{3\,\theta} = \eta_0\,\frac{c_3^2\,R}{\xi_{m,n}\,h_0}\,\frac{m}{r}\,J_m\left(\xi_{m,n}\frac{r}{R}\right)\,e^{i(m\theta\pm\Omega t)}. \tag{4.12}$$

As can be seen from Eqs. (4.11), (4.12) and (4.2), $v_{3\,r}$ and η are out-of-phase, while $v_{3\,\theta}$ and η are in phase. This implies that the first integral over time in Eq. (4.10) reduces to zero, whereas the second one, after taking real parts of complex exponentials, integrates to $\frac{1}{2}\,|v_{3\,\theta}|\,|\eta|$. Taking into account the results of the integration over time, from (4.10) we obtain the following expression:

$$\langle \Delta E \rangle = \frac{\rho\, m\, c_3^2}{\Omega\, h_0} \int_{r=0}^{R} r\, \mathrm{d}r\, \eta_0^2\, \frac{J_m^2\left(\xi_{m,n}\frac{r}{R}\right)}{r} \int_{\theta=0}^{2\pi} v_{v\,\theta}\, \mathrm{d}\theta. \tag{4.13}$$

Next, taking the stationary part of the surface displacement $\eta\,(r) = \eta_0\, J_m\left(\xi_{\mathrm{m,n}}\,\frac{r}{R}\right)$, we rewrite (4.13) as

$$\langle \Delta E \rangle = \frac{\rho\, m\, c_3^2}{\Omega\, h_0} \int_{r=0}^{R} \frac{\mathrm{d}r}{r}\, \eta^2\,(r) \int_{\theta=0}^{2\pi} v_{v\,\theta}\, r\, \mathrm{d}\theta. \tag{4.14}$$

Note that we moved r in (4.14) from the integral over r to the integral over θ. We now notice that the latter corresponds to a closed contour integral $\oint\, \vec{v}_v \cdot \mathrm{d}\vec{l}$, where the contour is a circle of radius r centered at the origin. We know that this closed contour integral defines the quantized vortex circulation around a loop encompassing the vortex core (Eq. (4.28)) and is equal to zero if the integration loop does not enclose the vortex core, and κ if it does. The transition between these two values of integration occurs for $r =$ offset, the radial offset of the point vortex from the origin. Therefore, we can rewrite Eq. (4.14) with a modified radial integration lower bound:

$$\langle \Delta E \rangle = \frac{\rho\, m\, c_3^2\, \kappa}{\Omega\, h_0} \int_{r=\mathrm{offset}}^{R} \frac{\mathrm{d}r}{r}\, \eta^2\,(r)\,. \tag{4.15}$$

Next, in order to link the kinetic energy difference (4.15) with the vortex-induced third-sound mode splitting, we notice that for a harmonic oscillator its E is proportional to Ω^2. Hence, $\frac{\Delta E}{E} = 2\frac{\Delta\Omega}{\Omega}$, and from here the splitting Δf (in Hz) equals

$$\Delta f = \frac{\Omega}{4\pi}\frac{\Delta E}{E}. \tag{4.16}$$

For rotationally noninvariant sound modes, i.e. modes with azimuthal order $m > 0$, the kinetic energy E is given by

$$E = \frac{1}{2}\int \rho\, \vec{v}^{\,2}\,(\vec{r})\, \mathrm{d}^3\,(\vec{r}) = \frac{\pi\, \rho\, c_3^2}{2\, h_0} \int_0^R \eta^2\,(r)\, r\, \mathrm{d}r. \tag{4.17}$$

Combining Eqs. (4.15) and (4.17), we arrive to the formula for the vortex-induced third-sound modes frequency splitting:

$$\Delta f = \frac{\kappa\, m}{2\,\pi^2}\, \frac{\int_{\mathrm{offset}}^{R} \frac{\mathrm{d}r}{r}\, \eta^2\,(r)}{\int_0^R \mathrm{d}r\, r\, \eta^2\,(r)}. \tag{4.18}$$

The obtained splitting applies to both superfluid helium thin films and Bose-Einstein condensates, with η being the film thickness perturbation for superfluid helium and density perturbation for BECs, respectively. We note that the splitting (4.18) does not depend on the superfluid parameters, such as film thickness or density, and that it is linear in vortex flow field (see Eq. (4.8)), such that the splitting obeys the superposition principle, whereby the splitting due to an ensemble of vortices is equal to the sum of the splittings induced by each individual vortex separately ($\Delta f_{\text{total}} = \sum_i \Delta f_i$). The result (4.18) can be interpreted such that the frequency splitting induced by a vortex offset from the disk origin by r_v is equal to the interaction energy between a vortex right at the disk origin ($r_v = 0$) and a third-sound wave in the area of the disk with radius beyond r_v.

Note that Eq. (4.18) does not diverge as the vortex offset tends to 0, as $\eta\,(0) = 0$ for all rotationally noninvariant ($m > 0$) Bessel modes. Remarkably, due to the contour-integral identity used in Eq. (4.14), the final result (4.18) does not require any knowledge of the vortex flow field $\vec{v}_v$ and depends only on the profile of the sound Bessel mode and the radial position of the vortex.

4.3 Finite-Element Modelling of Sound-Vortex Interactions

The interaction between persistent currents and sound modes has been analytically investigated and quantified for vortex distributions centered in circular domains, such as disk-shaped resonators [12, 40], in thin films of superfluid helium and for centered vortices in BECs confined in traps of various shapes [41, 42]. However, to date there has been no consistent approach to modelling a non-trivial vortex distribution in a non-trivial resonator shape, or for non-simply connected domains. Our approach to this modelling is to map the problem onto other areas of physics, namely map dynamics of superfluid sound modes onto hydrodynamics of an ideal gas and map vortex dynamics onto electrostatics. Using this method, we show how the interaction of an arbitrarily distributed ensemble of vortices with sound modes in any two-dimensional domain, perhaps multiply-connected, can be accurately modelled using the FEM-solver COMSOL ® Multiphysics 5.0.

In the following, we first introduce the aforementioned mapping of superfluid dynamics and vortex dynamics, and then present the results of FEM simulations of sound-vortex interactions. We also compare the FEM-results with the results of the analytical approach described earlier in this chapter. Tables, demonstrating the full mapping between quantities of superfluid dynamics, electrostatics, and acoustics of an ideal gas, can be found in Appendix A.1.

4.3.1 Third-Sound/Acoustics Correspondence

In Sect. 1.10 we provided the equations of motion describing the dynamics of third-sound waves (Eqs. (4.19) and (1.20)). Here we generalize those equations, so that they now can describe not only film deflections for third-sound waves but also density perturbations for Bose-Einstein condensates. In this case, equations describing superfluid hydrodynamics take the following form. The continuity equation, which derives from the mass conservation, reads [12]

$$\frac{d\rho}{dt} = -\vec{\nabla} \cdot (\rho \vec{v}). \tag{4.19}$$

And the Euler equation, which derives from momentum conservation, is given by

$$\frac{d\vec{v}}{dt} + (\vec{v} \cdot \vec{\nabla})\vec{v} = -g\vec{\nabla}\rho + C. \tag{4.20}$$

In the above equations, in the case of thin films of superfluid helium, $\vec{v}$ is the superfluid flow velocity, $\rho \rightarrow h$ is the film height, $g = \frac{3\alpha_{\text{vdw}}}{h^4}$ is the linearized van der Waals acceleration [27, 35], and $C = 0$. The hydrodynamics of Bose-Einstein condensates in the Thomas-Fermi limit at zero temperature [43, 44] can also be described by Eqs. (4.19) and (4.20) with $\vec{v}$ again being the flow velocity, ρ the BEC density, $g \rightarrow g_{\text{BEC}}/m^2$ the coupling strength, where m is the mass of an individual atom contributing to the condensate, and g_{BEC} is the atom-atom coupling. $C = -\vec{\nabla}U/m$ describes the trapping, where U is the extended potential [43] (see Table A.1 in Appendix A.1).

With an assumption of small density perturbations (BEC) or film height deflections (helium), η, from an equilibrium density/film height ρ_0, such as $\rho(\vec{r}, t) = \rho_0 + \eta(\vec{r}, t)$ with $\eta \ll \rho_0$, Eqs. (4.19) and (4.20) can be rewritten as

$$\dot{\eta} = -\rho_0\vec{\nabla} \cdot \vec{v} - \vec{v}\,\vec{\nabla}\eta, \tag{4.21}$$

and

$$\dot{\vec{v}} + (\vec{v} \cdot \vec{\nabla})\vec{v} = -g\vec{\nabla}\eta + C. \tag{4.22}$$

Next, we look at the equations of motion describing fluid dynamics of an ideal gas. The mass conservation equation is given by

$$\frac{d\rho}{dt} = -\vec{\nabla} \cdot (\rho \vec{u}), \tag{4.23}$$

and the equation for the momentum conservation reads [45]

$$\frac{d\vec{u}}{dt} + (\vec{u} \cdot \vec{\nabla})\vec{u} = -\frac{1}{\rho}\vec{\nabla}p. \tag{4.24}$$

In the above equations, $\vec{u}(\vec{r}, t)$ is the flow velocity, ρ is the gas density and p is the gas pressure. Condition of isentropic flow, i.e. when gas is in thermal equilibrium at all times, for an ideal gas entails the following [45]:

$$p = \gamma R T \rho \quad \text{and} \quad c^2 = \gamma R\, T. \tag{4.25}$$

Here, R is the specific gas constant, T is the gas temperature and c is the speed of sound. Next, we insert Eq. (4.25) in Eq. (4.24) and, in analogy with the superfluid dynamics equations, linearize for small density fluctuations, i.e. for $\rho(\vec{r}) = \rho_0 + \alpha(\vec{r})$ with $\alpha \ll \rho_0$:

$$\dot{\alpha} = -\rho_0 \vec{\nabla} \cdot \vec{u} - \vec{u}\, \vec{\nabla} \alpha \tag{4.26}$$

and

$$\dot{\vec{u}} + (\vec{u} \cdot \vec{\nabla})\, \vec{u} = -\frac{\gamma R\, T}{\rho_0} \vec{\nabla} \alpha. \tag{4.27}$$

Replacing $\alpha \to \eta$, $\vec{u} \to \vec{v}$ and $g \to c^2 = \gamma R T / \rho_0$, we end up with exactly same equations as (4.21) and (4.22), which describe superfluid dynamics in the limit of small amplitudes.

This allows us to model Eqs. (4.21) and (4.22) in COMSOL®. To do this, we utilize *Aeroacoustics → Linearized Euler, Frequency Domain(lef) module*, where appropriate boundary conditions are used (see Sect. 4.3.3). This provides us with a tool to determine sound eigenmodes within any arbitrarily-shaped bounded domain. Figure 4.3a shows examples of sound eigenmodes confined to circular and arbitrarily-shaped domains, with free ('*Neumann*') boundary conditions [31].

In this work, we study sound-vortex interactions in two-dimensional domains, as this reproduces experimental conditions described in Chap. 5 of this thesis, as well as those in Refs. [9, 12]. However, the presented FEM model can also be extended to three dimensions. The advantage of our FEM method is that it allows to implement a background flow field, which corresponds to the flow generated by an ensemble of vortices, and find the new sound eigenmodes in the presence of that background flow.

4.3.2 Vortex/Electrostatics Correspondence

As described in Sect. 1.7, a single vortex is characterized by a quantized circulation around a loop enclosing the vortex core [46, 47]:

$$\oint \vec{v}_v \cdot d\vec{l} = \kappa, \tag{4.28}$$

where $\kappa = h/m_{\text{He}}$ is the quantum of circulation, m_{He} is the mass of a helium atom, and $\vec{v}_v$ stands for the velocity field of the vortex flow. For a point vortex on a plane

$$\vec{v}_v(r) = \frac{\kappa}{2\pi r}\hat{e}_\theta, \tag{4.29}$$

where $\hat{e}_\theta$ is the tangential unit vector and r is the distance from the vortex core.

In order to map the vortex flow field onto electrostatics, we utilise the Gauss's law, which in two dimensions reads

$$\oint \vec{D} \cdot d\vec{n} = Q, \tag{4.30}$$

Here, $\vec{D}$ is the field of electric displacement and Q is the line charge. We next perform a rotation of the electric displacement field and replace it with the vortex flow field $\vec{v}_v$:

$$\begin{pmatrix} D_x \\ D_y \end{pmatrix} \rightarrow \begin{pmatrix} v_{v\,y} \\ -v_{v\,x} \end{pmatrix}. \tag{4.31}$$

By substituting $Q \rightarrow \kappa$, we arrive at the vortex flow field quantized circulation defined in Eq. (4.28), where $\vec{v}_v = v_{v\,x}\,\hat{e}_x + v_{v\,y}\,\hat{e}_y$. This manifests the mapping between vortex flow and electrostatics [46]. In the latter, a point charge is a source of divergence of the field of electric displacement $\vec{D}$. Upon the transformation given by (4.31), a point charge becomes a source of quantized circulation. Namely, the potential lines of $\vec{D}$ become streamlines of $\vec{v}_v$, and the streamlines of $\vec{D}$ become potential lines of $\vec{v}_v$, as shown in Fig. 4.3b.

In order to model these equations, we utilise *Electrostatics(es) module* of COMSOL®. This allows us to determine the vortex flow field on any arbitrary two-dimensional geometry. Figure 4.3b shows examples of vortex flow fields within a circular and an irregular geometries. Presence of vortices may significantly modify shapes of sound eigenmodes (Fig. 4.3c). The strength of the perturbation to sound modes depends on the number of vortices, their positions, and the resonator geometry. This method also allows us to model quantized circulation around a macroscopic topological defect in a multiply-connected domain, as described in Sect. 4.4.

4.3.3 Boundary Conditions

In order to solve differential equations of motion defining dynamics of superfluid sound waves, appropriate conditions have to be applied to the boundary of the domain which superfluid waves are confined to. We have already described these conditions in details in Sect. 1.10.2. In brief, two most common types of boundary conditions are distinguished as fixed (*'Dirichlet'*) and free (*'Neumann'*). Here we describe how these boundary conditions are implemented in our FEM method in order to model superfluid dynamics.

In COMSOL®, in order to find superfluid sound eigenmodes by solving an equivalent problem for an ideal gas, the free boundary condition is defined as a *rigid wall*,

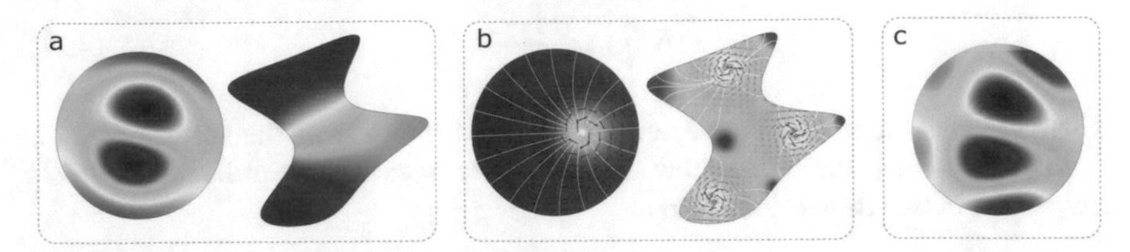

Fig. 4.3 **a** Finite Element Method (FEM) simulation of third-sound eigenmodes existing within a circular (left) and an arbitrarily-shaped (right) domain, with free boundary condition. Left: Bessel ($m = 1; n = 2$) mode. Right: lowest-frequency eigenmode of the geometry. Colour code indicates the magnitude of the displacement. **b** FEM modelling of the flow field generated by point vortices within bounded geometries shown in (**a**). Left: flow field of a clockwise (CW) point vortex displaced from the origin in a circular geometry (see Sect. 4.2.1). Right: flow field generated by two CW and one counter-clockwise (CCW) point vortices within an arbitrarily-shaped domain. Surface colour-code and red arrows represent the vortex flow velocity (in log scale). White lines indicate the streamlines of the unrotated electric displacement field $\vec{D}$, which are potential lines for the superfluid flow (see the electric-to-superfluid correspondence in Eq. (4.31)). **c** New 'perturbed' ($m = 1; n = 2$) third-sound eigenmode in the circular domain in the presence of the background flow generated by a large number of off-centered vortices

where volume is conserved and the gas pressure can oscillate freely at the boundary, whereas the fixed boundary condition corresponds to *fixed pressure*, where the gas pressure is fixed at the boundary and the gas can freely flow in and out of the domain [48].

Next, in order to choose an appropriate boundary condition for the vortex flow field, we note that the vortex flow is tangential to the boundary of the domain, $\vec{v}_v \cdot \vec{n} = 0$. After mapping the problem onto electrostatics, the vortex flow translates to electric field which is exactly perpendicular to the boundary, with no tangential component. To satisfy this condition, a perfect electric conductor should be chosen at the boundary. This can be realized by choosing the *ground*—boundary condition in COMSOL® [49]. Furthermore, the most appropriate boundary condition for modelling quantized circulation $n \times \kappa$ around an engineered topological defect in a resonator is the *floating potential* condition with a built-in charge of $Q = n \times \kappa$. This boundary condition ensures that the electric field is everywhere orthogonal to the boundary (due to the equal potential on the boundary) and satisfies the condition

$$\oint \vec{D} \cdot \vec{n}\, dl = Q. \tag{4.32}$$

By incorporating transformation (4.31) into (4.32), one can notice that this enforces superfluid flow to be always parallel to the boundary of the topological defect (i.e. no fluid inflow or outflow), and the quantized circulation condition is given by

$$\oint \vec{v} \cdot \mathrm{d}\vec{l} = n\,\kappa. \tag{4.33}$$

4.3.4 Sound-Vortex Interactions in a Circular Domain

We apply our FEM method to study sound-vortex interactions in a circular domain. Namely, we investigate the experimentally relevant example of the interaction of a single quantized vortex with third-sound modes in a disk-shaped resonator with free (‘*Neumann*’) boundary condition for superfluid waves. This analysis is applicable to the geometries used in our experiments with a microtoroidal optomechanical resonator of $R \sim 30$ μm radius coated with thin film of superfluid helium [29, 30] (see Fig. 4.4a and Chap. 5 for experimental details). This study is also relevant for the experiments from Refs. [5, 11, 12, 50], as well as for the experiments with two-dimensional Bose-Einstein condensates confined by a hard-walled trap [9].

The mechanism of sound-vortex coupling is elaborated in Sect. 4.2 (also see Fig. 4.4b). Here we investigate how the coupling is affected when the vortex position on a circular resonator is varied. As the radial offset r_v of the vortex from the disk center increases, the splitting reduces, as can be seen in Fig. 4.4d. Firstly, this is a consequence of the lower total kinetic energy of the vortex flow as the vortex core approaches the boundary. The closer the vortex nears the disk boundary, the more of its flow field is clipped, resulting in a lower total kinetic energy of the flow. Secondly, given the vortex flow velocity distribution, the highest kinetic energy of the vortex flow is concentrated around the vortex core. Thus, symmetry dictates that the overlap—and hence sound frequency splitting—between vortex and sound flow fields is maximized for a vortex at the disk centre ($r_v = 0$). As an example, frequency shifts of the free-boundary-condition CW and CCW ($m = 1$, $n = 8$) sound Bessel modes are plotted as a function of the vortex radial offset r_v in Fig. 4.4c, while the splitting dependence of this and a few other third-sound modes on r_v is shown in Fig. 4.4d. As can be seen from this figure, modes with different azimuthal number m experience substantially different dependence of their splittings on r_v. And Eq. (4.18) contains an explanation for this difference. As discussed at the end of Sect. 4.2.3, only that fraction of the sound wave that is confined beyond the vortex radial position r_v contributes to the splitting (an illustration of this is in the inset in Fig. 4.4e). And since most of the kinetic energy of the modes with (low m, high n) order is located close to the center of the disk, splitting per vortex experienced by these modes decays rapidly with r_v (modes $(1, 3)$, $(1, 5)$, $(1, 8)$ in Fig. 4.4d). While modes with (high m, low n) order have more radially extended kinetic energy distribution and, thus, sustain significant splitting at higher radii (mode $(5, 4)$ in Fig. 4.4d). Each point in Fig. 4.4c, d corresponds to one result of the FEM simulation, as the vortex is stepped outwards from the disk center. As the vortex reaches the boundary of the domain, the splitting vanishes.

As we have now established how to compute the vortex-induced third-sound modes splitting with both analytical analysis and FEM method, we are in position to compare the results of these two approaches. We find that the two methods agree within 10% of the maximal splitting at zero radial offset of the vortex from the disk origin ($r_v = 0$). The remaining difference is plotted in Fig. 4.5. We attribute this difference to an alteration of the third-sound mode shape induced by the nonlinear

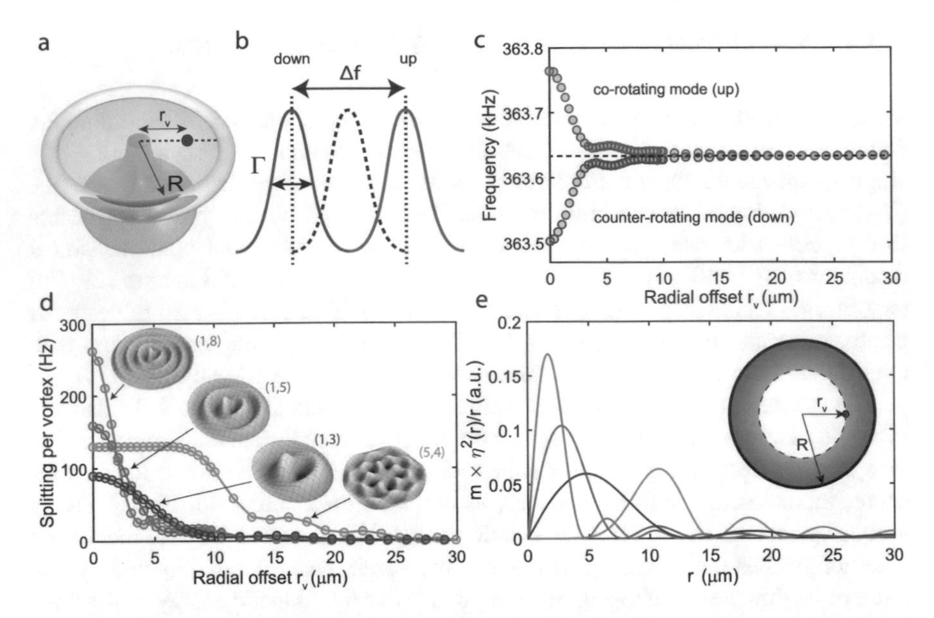

Fig. 4.4 **a** Illustration of a microtoroidal resonator of radius $R = 30$ μm coated with thin film of superfluid helium, with one quantized vortex offset from the disk origin by r_v. The red dot indicates the vortex core. **b** Schematic representation of a vortex-induced third-sound mode splitting. Mode depicted by the dashed line is a degenerate third-sound mode in the vortex absence. Blue and red curves respectively represent frequency up-shifted and down-shifted modes in the presence of a vortex which lifts the degeneracy (see Fig. 4.1b). Splitting Δf is proportional to the number of vortices present in the system and depends on a spatial position of the vortex on a disk (see (**a**)). Γ indicates modes linewidth. **c** Frequency of the co- and counter-rotating ($m = 1$, $n = 8$) -Bessel mode, with one quantized vortex present on the resonator, as a function of the vortex offset r_v from the disk origin. The frequency up-shifted (down-shifted) mode is shown in blue (red) and corresponds to the right (left) peak in (**b**). **d** Frequency splitting dependence of (1,3), (1,5), (1,8), and (5,4) third-sound modes on the radial offset of the vortex from the disk origin. Spatial profiles of the modes are shown as insets. **e** Displacement amplitudes of modes shown in (**d**) as a function of radius. Inset: annular region cut out by a circle—with its radius defined by the vortex position r_v—whose interaction with the vortex accounts for the majority of the splitting (see text). All plots use free boundary condition for the modes; each dot in (**c**), (**d**) and (**e**) represents a result of the FEM-simulation

convective $(\vec{v} \cdot \vec{\nabla})\vec{v}$ term in Eq. (4.22):

$$(\vec{v} \cdot \vec{\nabla})\vec{v} = (\vec{v}_v \cdot \vec{\nabla})\vec{v}_v + (\vec{v}_s \cdot \vec{\nabla})\vec{v}_s + (\vec{v}_v \cdot \vec{\nabla})\vec{v}_s + (\vec{v}_s \cdot \vec{\nabla})\vec{v}_v. \tag{4.34}$$

The analytical perturbation approach neglects all of the above terms. Whereas the FEM model neglects only $(\vec{v}_s \cdot \vec{\nabla})\vec{v}_s$, which is a requirement to obtain sound eigenmodes. The latter approximation is valid in the quasi-static limit, where the third-sound flow velocity is small compared to the vortex flow velocity.

As we mentioned in Sect. 4.2.3, the total splitting is subject to linear superposition of splittings, i.e. the total splitting is obtained by summing splittings induced

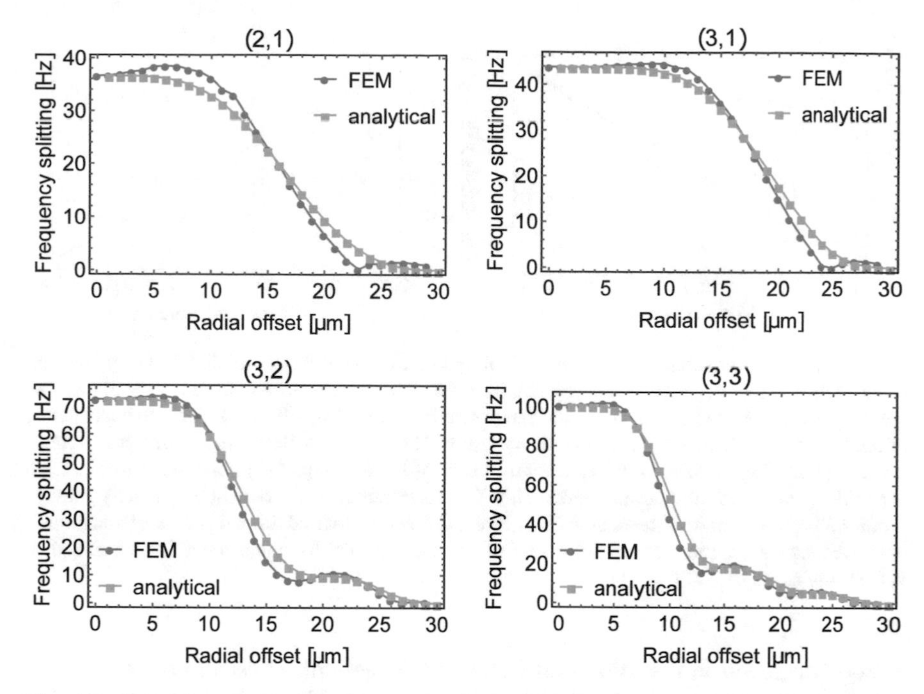

Fig. 4.5 Comparison between third-sound modes splittings per vortex obtained with the FEM method and the analytical perturbative approach (Eq. (4.18)) for four different Bessel modes labelled by their (m, n) order. The results show a good agreement (without any scaling parameter) between both methods within 10% of the maximal splitting at zero vortex radial offset ($r_v = 0$). The plots are obtained for the resonator with radius $R = 30\ \mu$m, and with fixed boundary condition for the third-sound modes. Note, that some small quantitative discrepancies between results obtained with both methods still remain. For example, from Eq. (4.18), the analytical splitting has to be a monotonically decreasing function of the vortex radial offset, while the FEM solution features some regions of increased splitting with radial offset. We attribute these discrepancies to vortex-induced changes in third-sound eigenmode shapes (see Fig. 4.3c), which are not accounted for in the perturbative analytical approach

by individual vortices separately. Here we verify this superposition with the FEM simulations, where linearity is generally maintained up to large vortex charges on the order of $\sim 10^2\ \kappa$, as shown in Fig. 4.6a.

4.4 Proposed Experimental Scheme for Detection of Single Quanta of Circulation in Two-Dimensional Superfluid Helium

Quantized vortices and their interactions with other superfluid elementary excitations determine dynamics of superfluids. Remarkably, while quantized vortices are crucial to our understanding of superfluidity in two-dimensions, they have never been

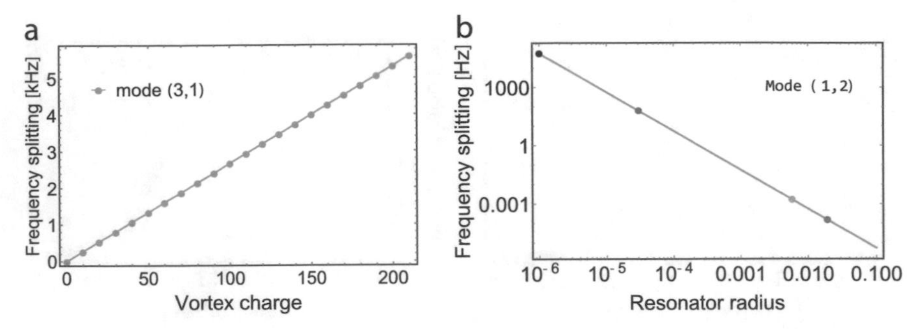

Fig. 4.6 **a** FEM computation of the ($m = 3$, $n = 1$) third-sound Bessel mode frequency splitting induced by a vortex centered at the disk origin, whose charge is increased from κ to $> 200\kappa$. The splitting displays linearity over the entire range of the vortex charge. **b** Resonator-size dependence of splitting per centered vortex for the ($m = 1$, $n = 2$) third-sound Bessel mode with free boundary condition. Experimental devices shown in red [50] and orange [11] correspond to cm-scale capacitively detected third-sound waves. Blue dot corresponds to the optical whispering-gallery-mode (WGM) microtoroid resonator [29]. Black dot shows two additional orders of magnitude improvement over current state-of-the-art that can be achieved by going towards micron-radius WGM resonators [51, 52]

directly observed in two-dimensional superfluid helium. Physical parameters and properties of thin superfluid helium films present a significant challenge to detecting single quanta of circulation in this quantum fluid. Here are a few of the main challenges. First, the normal-fluid core of a vortex in superfluid helium-4 is roughly one Angström in diameter [35]. Second, a typical thickness of a superfluid helium film is less than 20 nm, and the refractive index of liquid helium is close to that of vacuum ($n_{\mathrm{He}} \approx 1.029$). Combined, these characteristics preclude direct optical imaging, as can be performed in Bose-Einstein condensates [33, 53]. In bulk superfluid helium, many imaging techniques have utilised various kinds of tracer particles [54–56], such as micrometer-sized frozen hydrogen crystals or electron bubbles. The tracer particles scatter light and are pulled in to the vortex core, enabling, for instance, the recent observation of Kelvin waves [56] in bulk helium. Naturally, such an approach is significantly more difficult in two-dimensional films due to their few-nanometre thickness.

Here we study the requirements for observing single quantized vortices in two-dimensional superfluid helium and propose an experimental scheme for the detection of single quanta of circulation. The quantization of circulation can be observed via the shift in third-sound frequencies induced by an addition/subtraction of a single vortex.

There are a few challenges that must be addressed in order for the vortex-induced frequency splitting Δf (Eq. (4.18)) experienced by a third-sound wave [11, 12] to reveal quantized steps. First, in order to be resolvable, the vortex-induced splitting must be greater than the linewidth of the third-sound resonances, i.e. $\Delta f \gtrsim \Gamma$, as shown in Fig. 4.4b. However, we note that this is not a strict condition. Using certain techniques, the splitting could be resolved even if $\Delta f < \Gamma$. Second, any motion of vortices on the confining surface of the resonator will lead to a continuous evolution of the splitting due to the continuous nature of the splitting function $\Delta f(r_v)$ (see Fig. 4.4d). Hence, the vortex motion would mask the quantization of circulation and preclude the observation of discrete steps in the vortex-induced splitting. Third, vortex nucleation or annihilation events at different radial coordinates on the disk surface may result in unequal step sizes, making it difficult to interpret the observed splitting.

The first challenge can be addressed by designing resonators that are sufficiently small, and by controlling dissipation mechanisms in these devices. This can be achieved, for example, by engineering a resonator with an atomically smooth surface and that is decoupled from its ambient environment by a small anchor point [29] which would allow losses to be minimized. Alternatively, an appropriate engineering of the substrate material [57] would also enable reduced dissipation. In the experimental work presented in Chap. 5 of this thesis the frequency splittings Δf per vortex are comparable to linewidths Γ of the observed third-sound resonances [31]. A solution to the second and third challenges is to spatially constrain the position of the circulation around a macroscopic topological defect engineered on the surface of the resonator. For example, if we replace the topological defect naturally formed by the normal fluid core (of radius a_0) of a superfluid vortex by a microfabricated hole of radius $R \gg a_0$, the maximal flow velocity due to the quantized circulation becomes $\frac{\kappa}{2\pi R} \ll \frac{\kappa}{2\pi a0}$ (see Eq. (4.28)). This effectively clips the high velocity region of the flow and is thus energetically favourable. The circulation will then preferentially accumulate around this manufactured defect and can reach large values in terms of circulation quanta $\kappa = h/m_{\mathrm{He}}$, as has been observed in the experiments with bulk helium, whereby it has been spun up in an annular container [58, 59]. The quantization of the circulation then manifests as discrete steps in the splitting experienced by third-sound modes confined to the surface of the resonator.

This approach can be deemed as a two-dimensional analog of Vinen's experimental technique which he exploited for the first observation of circulation quanta in bulk helium [60, 61]. In those experiments circulation locked to a vibrating wire lifted the degeneracy between the wire's normal modes of vibration. In Fig. 4.7 we propose a practical realization of such a device which is based on a circular whispering-gallery-mode geometry, as used in our other works described in Chaps. 2, 3 and 5 of this thesis and in Refs. [29–31]. We leverage advantages of the FEM simulation to design a domain in which splitting Δf is maximised. Such a device features a single-spoked annular geometry [62] (Fig. 4.7c).

We calculate the superfluid flow field resulting from quantized circulation about the central topological defect, as shown in Fig. 4.7c. Without loss of generality, we assume that there are no free vortex cores elsewhere in the domain and the circulation locked around the defect is the only source of circulation on the structure. The flow field is calculated through FEM simulation with the use of the '*floating potential*' boundary condition for the inner boundary, which enforces both the prescribed circulation strength and parallelism of the flow to the boundary (see Sect. 4.3.3). The results are shown in Fig. 4.8a (left). The superfluid flow is predominantly confined to the outer annulus. Flows up and down the spoke and inside the central disk are negligible. This can be understood by considering closed contours 1 and 2 both

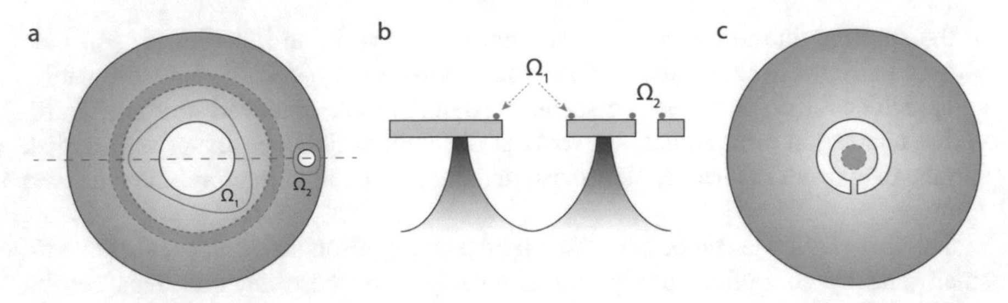

Fig. 4.7 Proposed device for detection of a single quanta of circulation. Topological defect is engineered within a disk-shaped resonator geometry. **a** Top-view of an annular-shaped resonator where superfluid circulation accumulates around the central hole [38]. Blue-shaded area represents the disk, while the dashed grey region depicts the device's pedestal which the disk rests upon. Red contours Ω_1 and Ω_2 represent closed loops around holes in the resonator. **b** Cross-section of the resonator through the dashed green line in (**a**), showing how contour Ω_1 can be continuously deformed and collapsed, while contour Ω_2 cannot. Hence, contour Ω_2 encloses a real topological defect. **c** Single-spoked annular disk geometry [62], in which the central hole is topologically identical to that enclosed by contour Ω_2 in (**b**)

enclosing the central hole (Fig. 4.8a(right)). The circulation around both contours must be equal (Eq. (4.28)), which implies that the additional circulation along the extra path contained in contour 1 is negligible.

Figure 4.8b shows an example of a third-sound mode which can exist on the proposed spoked resonator. As radius of the central defect asymptotically reduces to zero, the mode in Fig. 4.8b becomes the ($m = 1$, $n = 2$) Bessel eigenmode of a circular resonator. In the spoked geometry, the presence of the spoke lifts the degeneracy between the two normal modes. This is a *geometrical* splitting native to the structure even in the absence of circulation. This geometrical splitting can be adjusted by engineering the spoke with required geometrical parameters. The normal mode that has a stronger interaction with the spoke (bottom in Fig. 4.8b) feels an effectively larger resonator and, thus, has a lower resonance frequency.

In Fig. 4.8c we demonstrate how this native geometrical splitting [11] impacts the third-sound mode splitting as a function of the magnitude of circulation pinned around the central defect. Each black dot in the figure represents a finite-element simulation of the splitting between the high and low frequency eigenmodes shown in Fig. 4.8b. The splitting is computed as a function of the number of circulation quanta around the central hole. The total splitting can also be calculated analytically with an expression of the form [37]

$$s_{\text{total}} = \sqrt{s_{\text{circ}}^2 + s_{\text{geo}}^2}, \tag{4.35}$$

where $s_{\text{geo}} = 700$ Hz is the native geometrical splitting for the proposed device (dashed orange line in Fig. 4.8c), $s_{\text{circ}} = N \times s_{\text{circ},0}$ is the total circulation-induced splitting where N is the number of circulation quanta and $s_{\text{circ},0}$ is the splitting per quantum. The solid red line represents s_{total}, as given by Eq. (4.35). Figure 4.8d

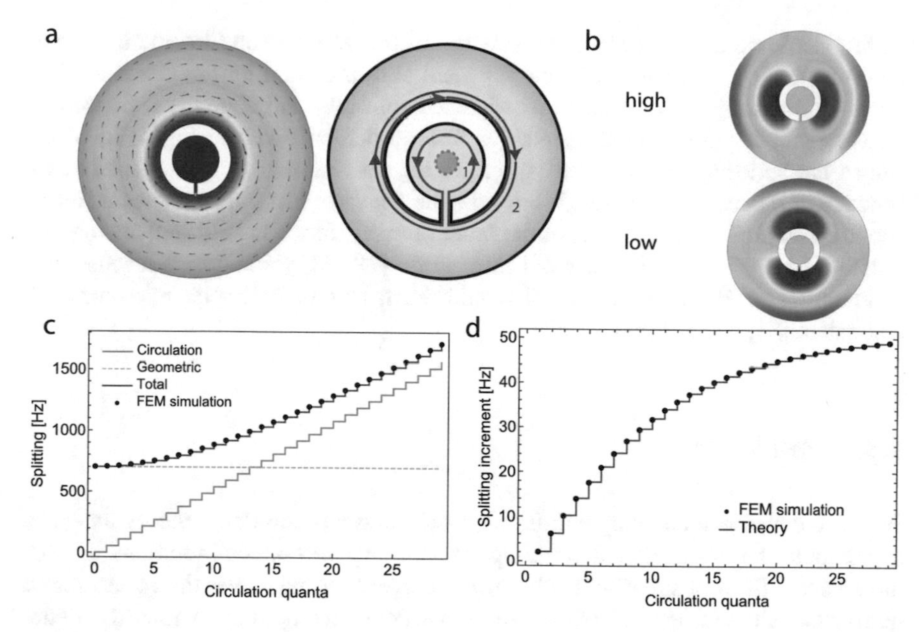

Fig. 4.8 **a** Left: superfluid flow field resulting from quantized circulation pinned around the central topological defect (red arrows). No free vortex cores are present in the domain. Colour code represents the magnitude of the flow velocity (red—fast; blue—slow). Right: contours 1 (blue) and 2 (red) both enclose the central defect and have therefore identical circulation (Eq. (4.28)). Dimensions of the resonator used for simulation: outer radius = 20 μm; inner radius = 4 μm; slot width = 2 μm; spoke width = 0.5 μm. **b** Non-degenerate eigenmodes of the spoked resonator, split by the presence of the spoke, with free boundary conditions at both boundaries. Colour code represents the magnitude of the surface displacement. **c** Contribution to the total mode splitting (red curve, see Eq. (4.35)) from geometrical splitting (dashed orange, $s_{\text{geo}} = 700$ Hz) and circulation-induced splitting (blue curve, $s_{\text{circ}} = N \times s_{\text{circ},0} = N \times 54$ Hz). Splitting is shown as a function of the number of circulation quanta pinned around the topological defect. **d** Splitting increment per added quantum of circulation

shows the splitting increment, induced by each additional quantum of circulation, as a function of the number of circulation quanta already present around the defect. This splitting increment is an experimentally relevant parameter. This demonstrates that the geometrical splitting induced by the spoke (or any unwanted deviation from circularity) will mask the influence of the circulation-induced splitting for small values of the circulation quanta, and, hence, reduce the visibility of the discrete steps. For larger values of the circulation present around the defect, the size of the steps will asymptote towards the value $s_{\text{circ},0} = 54$ Hz. This occurs as the normal mode basis (see Fig. 4.8b) gradually transitions from orthogonal standing waves to counter-propagating waves. Figure 4.8d demonstrates that the proposed device would yield quantized steps in the third-sound mode splitting on the order of 50 Hz, which is within reach of our current experimental resolution [29, 31] (see Chap. 5).

We also note also that since the strength of the vortex-sound interaction scales inversely with the vortex-sound confinement area provided by the resonator (see Sect. 4.2, the magnitude of the vortex-indiced splitting can be greatly enhanced by utilising smaller third-sound resonators. This is demonstrated in Fig. 4.6b, which shows the splitting per centered vortex on the $(m = 1, n = 2)$ Bessel mode as a function of the resonator radius. While with the cm-scale capacitively detected third-sound resonators the splitting is of the order of milli-Hertz [11, 50] (red and orange dots in Fig. 4.6), with microtoroidal resonators [29, 31] (blue dot) it reaches tens to hundreds of Hz. Furthermore, it would reach tens of kHz with micron-radius resonators [51, 52, 63] (black dot).

4.5 Conclusion

We have developed an analytical framework, using a perturbation theory analysis, to compute the vortex-sound coupling for arbitrary vortex configurations in two-dimensional films of superfluid helium. The derived formula gives the vortex-sound interaction rate at the level of a single vortex and does not require any knowledge of the vortex flow field. The perturbation theory analysis is performed for the analytically tractable case of a circular resonator geometry.

Furthermore, we have developed finite-element modelling tools to compute the interaction between any vortex flow and any sound wave in arbitrary, and potentially multiply-connected, two-dimensional domains. The developed finite-element method is versatile and applicable to both BEC superfluids and superfluid helium thin films. Versatile numerical techniques are needed for modelling vortex and sound velocity fields, as well as their interactions. Analytical solutions for the vortex flow field are only available for the cases when the domain features a high degree of symmetry. However, even if such solutions exist, any deviations from simple geometries like a disk make the implementation of the method of images, required for the cancellation of the normal component of the vortex flow on the resonator boundary [39], increasingly challenging. For multiply-connected domains, solutions often require an infinite series of images as the domain possesses two or more boundaries, and analytical solutions only exist for simple limit-cases such as a centered annular domain [59]. In case of nontrivial geometries one needs to rely on conformal mapping techniques [9, 64]. We compare the results of the developed finite-element method to our analytical framework for the vortex-sound coupling and verify the validity of our numerical approach.

Understanding the physics of strongly interacting superfluids requires precise knowledge of how quantized vortices in superfluids and persistent currents couple to sound waves at the level of a single vortex or circulation quantum. The modelling techniques presented in this work may help shed light on the validity of phenomenological models such as the point-vortex model [7, 65] in superfluids, as well as advance our understanding of quantum turbulence [10, 24–26] and energy dissipation in superfluids [5, 6, 66].

References

1. Berezinskii VL (1971) Destruction of long-range order in one-dimensional and two-dimensional systems having a continuous symmetry group I. Classical systems. Sov J Exp Theor Phys 32(3):493–500
2. Berezinskii VL (1972) Destruction of long-range order in one-dimensional and two-dimensional systems possessing a continuous symmetry group. II. Quantum systems. Sov J Exp Theor Phys 34(3):610–616
3. Kosterlitz JM, Thouless DJ (1972) Long range order and metastability in two dimensional solids and superfluids. (Application of dislocation theory). J Phys C Solid State Phys 5(11):L124
4. Kosterlitz JM, Thouless DJ (1973) Ordering, metastability and phase transitions in two-dimensional systems. J Phys C Solid State Phys 6(7):1181
5. Hoffmann JA, Penanen K, Davis JC, Packard RE (2004) Measurements of attenuation of third sound: evidence of trapped vorticity in thick films of superfluid ^{4}He. J Low Temp Phys 135(3–4):177–202
6. Penanen K, Packard RE (2002) A model for third sound attenuation in thick ^{4}He films. J Low Temp Phys 128(1–2):25–35
7. Simula T, Davis MJ, Helmerson K (2014) Emergence of order from turbulence in an isolated planar superfluid. Phys Rev Lett 113(16):165302
8. Onsager L (1949) Statistical hydrodynamics. Nuovo Cimento Suppl 6(S2):279–287
9. Gauthier G, Reeves MT, Yu X, Bradley AS, Baker MA, Bell TA, Rubinsztein-Dunlop H, Davis MJ, Neely TW (2019) Giant vortex clusters in a two-dimensional quantum fluid. Science 364(6447):1264–1267
10. Johnstone SP, Groszek AJ, Starkey PT, Billington CJ, Simula TP, Helmerson K (2019) Evolution of large-scale flow from turbulence in a two-dimensional superfluid. Science 364(6447):1267–1271
11. Ellis FM, Luo H (1989) Observation of the persistent-current splitting of a third-sound resonator. Phys Rev B 39(4):2703–2706
12. Ellis FM, Li L (1993) Quantum swirling of superfluid helium films. Phys Rev Lett 71(10):1577–1580
13. Parker N, Proukakis NP, Barenghi CF, Adams CS (2004) Controlled vortex-sound interactions in atomic Bose-Einstein condensates. Phys Rev Lett 98(16):160403
14. Sonin EB (1997) Magnus force in superfluids and superconductors. Phys Rev B 55(1):485–501
15. Bergman D (1969) Hydrodynamics and third sound in thin He II films. Phys Rev 188(1):370–384
16. Bergman DJ (1971) Third sound in superfluid helium films of arbitrary thickness. Phys Rev A 3(6):2058–2066
17. Berthold JE, Bishop DJ, Reppy JD (1977) Superfluid transition of ^{4}He films adsorbed on porous vycor glass. Phys Rev Lett 39(6):348–352
18. Bishop DJ, Reppy JD (1978) Study of the superfluid transition in two-dimensional He 4 films. Phys Rev Lett 40(26):1727
19. Hadzibabic Z, Krüger P, Cheneau M, Battelier B, Dalibard J (2006) Berezinskii–Kosterlitz–Thouless crossover in a trapped atomic gas. Nature 441(7097):1118–1121
20. Holzmann M, Baym G, Blaizot J-P, Laloë F (2007) Superfluid transition of homogeneous and trapped two-dimensional Bose gases. Proc Natl Acad Sci 104(5):1476–1481
21. Qi X-L, Hughes TL, Raghu S, Zhang S-C (2009) Time-reversal-invariant topological superconductors and superfluids in two and three dimensions. Phys Rev Lett 102(18):187001
22. Lagoudakis KG, Wouters M, Richard M, Baas A, Carusotto I, André R, Dang LS, Deveaud-Plédran B (2008) Quantized vortices in an exciton–polariton condensate. Nat Phys 4(9):706–710
23. Roumpos G, Fraser MD, Löffler A, Höfling S, Forchel A, Yamamoto Y (2011) Single vortex–antivortex pair in an exciton-polariton condensate. Nat Phys 7(2):129–133
24. Barenghi CF, Skrbek L, Sreenivasan KR (2014) Introduction to quantum turbulence. Proc Natl Acad Sci 111:4647–4652

25. Paoletti MS, Lathrop DP (2011) Quantum turbulence. Annu Rev Condens Matter Phys 2(1):213–234
26. Skrbek L (2011) Quantum turbulence. J Phys Conf Ser 318(1):012004
27. Atkins KR (1959) Third and fourth sound in liquid Helium II. Phys Rev 113(4):962–965
28. Ville JL, Saint-Jalm R, Le Cerf É, Aidelsburger M, Nascimbène S, Dalibard J, Beugnon J (2018) Sound propagation in a uniform superfluid two-dimensional Bose gas. Phys Rev Lett 121(14):145301
29. Harris GI, McAuslan DL, Sheridan E, Sachkou Y, Baker C, Bowen WP (2016) Laser cooling and control of excitations in superfluid helium. Nat Phys 12(8):788–793
30. McAuslan DL, Harris GI, Baker C, Sachkou Y, He X, Sheridan E, Bowen WP (2016) Microphotonic forces from superfluid flow. Phys Rev X 6(2):021012
31. Sachkou YP, Baker CG, Harris GI, Stockdale OR, Forstner S, Reeves MT, He X, McAuslan DL, Bradley AS, Davis MJ, Bowen WP (2019) Coherent vortex dynamics in a strongly interacting superfluid on a silicon chip. Science 366(6472):1480–1485
32. Matthews MR, Anderson BP, Haljan PC, Hall DS, Wieman CE, Cornell EA (1999) Vortices in a Bose-Einstein condensate. Phys Rev Lett 83(13):2498–2501
33. Weiler CN, Neely TW, Scherer DR, Bradley AS, Davis MJ, Anderson BP (2008) Spontaneous vortices in the formation of Bose-Einstein condensates. Nature 455(7215):948–951
34. Freilich DV, Bianchi DM, Kaufman AM, Langin TK, Hall DS (2010) Real-time dynamics of single vortex lines and vortex dipoles in a Bose-Einstein condensate. Science 329(5996):1182–1185
35. Tilley DR, Tilley J (1990) Superfluidity and superconductivity. CRC Press
36. Wilson C, Ellis FM (1995) Vortex creation and pinning in high amplitude third sound waves. J Low Temp Phys 101(3):507–512
37. Ellis FM, Wilson CL (1998) Excitation and relaxation of film flow induced by third sound. J Low Temp Phys 113(3–4):411–416
38. Baker CG, Harris GI, McAuslan DL, Sachkou Y, He X, Bowen WP (2016) Theoretical framework for thin film superfluid optomechanics: towards the quantum regime. New J Phys 18(12):123025
39. Lamb H (1993) Hydrodynamics. Cambridge University Press
40. Wilson CL (1998) Swirling superfluid ^{4}He films. PhD thesis, Wesleyan University
41. Zambelli F, Stringari S (1998) Quantized vortices and collective oscillations of a trapped Bose-Einstein condensate. Phys Rev Lett 81(9):1754–1757
42. Svidzinsky AA, Fetter AL (1998) Normal modes of a vortex in a trapped Bose-Einstein condensate. Phys Rev A 58(4):3168–3179
43. Schuck P, Viñas X (2000) Thomas-Fermi approximation for Bose-Einstein condensates in traps. Phys Rev A 61(4):043603
44. Pitaevskii L, Stringari S (2016) Bose-Einstein condensation and superfluidity. Oxford University Press
45. Rienstra SW, Hirschberg A (2004) An introduction to acoustics. Technische Universiteit Eindhoven
46. Ambegaokar V, Halperin BI, Nelson DR, Siggia ED (1980) Dynamics of superfluid films. Phys Rev B 21(5):1806–1826
47. Donnelly RJ (1991) Quantized vortices in Helium II. Cambridge University Press
48. Acoustics Module User's Guide, version 5.0, COMSOL, Inc
49. AC/DC Module User's Guide, version 5.0, COMSOL, Inc
50. Schechter AMR, Simmonds RW, Packard RE, Davis JC (1998) Observation of 'third sound' in superfluid He-3. Nature 396(6711):554–557
51. Gil-Santos E, Baker C, Lemaître A, Gomez C, Leo G, Favero I (2017) Scalable high-precision tuning of photonic resonators by resonant cavity-enhanced photoelectrochemical etching. Nat Commun 8:14267
52. Gil-Santos E, Baker C, Nguyen DT, Hease W, Gomez C, Lemaître A, Ducci S, Leo G, Favero I (2015) High-frequency nano-optomechanical disk resonators in liquids. Nat Nanotechnol 10(9):810–816

53. Wilson KE, Newman ZL, Lowney JD, Anderson BP (2015) In situ imaging of vortices in Bose-Einstein condensates. Phys Rev A 91(2):023621
54. Yarmchuk EJ, Gordon MJV, Packard RE (1979) Observation of stationary vortex arrays in rotating superfluid helium. Phys Rev Lett 43(3):214–217
55. Bewley GP, Lathrop DP, Sreenivasan KR (2006) Superfluid helium: visualization of quantized vortices. Nature 441(7093):588
56. Fonda E, Meichle DP, Ouellette NT, Hormoz S, Lathrop DP (2014) Direct observation of Kelvin waves excited by quantized vortex reconnection. Proc Natl Acad Sci 111:4707–4710
57. Souris F, Rojas X, Kim PH, Davis JP (2017) Ultralow-dissipation superfluid micromechanical resonator. Phys Rev Appl 7(4):044008
58. Bendt PJ (1967) Superfluid flow transitions in rotating narrow annuli. Phys Rev 164(1):262–267
59. Fetter AL (1967) Low-lying superfluid states in a rotating annulus. Phys Rev 153(1):285–296
60. Vinen WF (1961) The detection of single quanta of circulation in liquid Helium II. Proc R Soc A Math Phys Eng Sci 260(1301):218–236
61. Whitmore SC, Zimmermann W Jr (1968) Observation of quantized circulation in superfluid helium. Phys Rev 166(1):181
62. Baker CG, Bekker C, McAuslan DL, Sheridan E, Bowen WP (2016) High bandwidth on-chip capacitive tuning of microtoroid resonators. Opt Express 24(18):20400
63. Baker C, Hease W, Nguyen D-T, Andronico A, Ducci S, Leo G, Favero I (2014) Photoelastic coupling in gallium arsenide optomechanical disk resonators. Opt Express 22(12):14072–14086
64. Saffman PG (1992) Vortex dynamics. Cambridge University Press
65. Billam T, Reeves M, Bradley A (2015) Spectral energy transport in two-dimensional quantum vortex dynamics. Phys Rev A 91(2):023615
66. Ekholm DT, Hallock RB (1980) Studies of the decay of persistent currents in unsaturated films of superfluid ^{4}He. Phys Rev B 21(9):3902–3912

Chapter 5
Observation of Coherent Vortex Dynamics in Two-Dimensional Superfluid Helium

Two-dimensional superfluidity and quantum turbulence are emergent phenomena native to various condensed matter systems ranging from ultracold quantum gases to superfluid helium. However, to date, there is no complete microscopic model for these phenomena in strongly interacting systems. Therefore, experiments are crucial for understanding of their rich behaviour. Both two-dimensional superfluidity and quantum turbulence are directly connected to the microscopic dynamics of quantized vortices. Yet, strong surface effects have prevented direct observations of coherent vortex dynamics in strongly-interacting two-dimensional systems. In this work we overcome this long-standing challenge by confining a two-dimensional droplet of superfluid helium to the atomically-smooth surface of a on-chip optical microcavity. This allows laser-initiation of vortex clusters and nondestructive observation of their decay in a single shot. The smooth surface of the chip and low superfluid temperature, comparing to the temperature of the Berezinskii-Kosterlitz-Thouless phase transition, enable a five orders-of-magnitude reduction in thermal vortex diffusion. Thus, we observe that coherent vortex dynamics occurs at a rate that is more than five orders of magnitude faster than dissipation. Our work establishes a new on-chip platform to study emergent phenomena in strongly interacting superfluids, construct quantum technologies such as precision inertial sensors, and even potentially test astrophysical dynamics, such as those in the superfluid core of neutron stars, in the laboratory.[1]

[1]This chapter is reproduced with some modification from: **Y. P. Sachkou**, C. G. Baker, G. I. Harris, O. R. Stockdale, S. Forstner, M. T. Reeves, X. He, D. L. McAuslan, A. S. Bradley, M. J. Davis, and W. P. Bowen. Coherent vortex dynamics in a strongly interacting superfluid on a silicon chip. *Science* 366, 1480–1485, 2019. Reprinted with permission from AAAS. Readers may view, browse, and/or download material for temporary copying purposes only, provided these uses are for noncommercial personal purposes. Except as provided by law, this material may not be further reproduced, distributed, transmitted, modified, adapted, performed, displayed, published, or sold in whole or in part, without prior written permission from the publisher.

Y. Sachkou, *Probing Two-Dimensional Quantum Fluids with Cavity Optomechanics*, Springer Theses, https://doi.org/10.1007/978-3-030-52766-2_5

5.1 Introduction

Strongly interacting many-body quantum systems exhibit rich behaviours and play important roles in various areas of physics ranging from superconductivity [1] to quantum simulation and computation [2, 3]. They are also of a great significance to such areas as neutron star dynamics [4, 5], quark-gluon plasmas in the early universe [6], and even string theory [7]. Superfluidity in helium was the first example of such unusual behaviour of strongly interacting quantum many-body systems. Remarkably, it was found to persist even in thin two-dimensional films [8], for which the Mermin-Wagner-Hohenberg theorem precludes condensation into a superfluid phase in the thermodynamic limit [9]. This apparent contradiction was resolved by Berezinskii, Kosterlitz and Thouless (BKT), who predicted that quantized vortices allow a topological phase transition into superfluidity [10, 11]. It is now recognized that quantized vortices also dominate much of the out-of-equilibrium dynamics of two-dimensional superfluids, such as quantum turbulence [12].

Recently, the demonstration of laser control and imaging of vortices in ultracold gases [13, 14] and semiconductor exciton-polariton systems [15, 16] has enabled extensive capabilities to study superfluid dynamics [17] including, for example, the formation of collective vortex dipoles with negative temperature and large-scale order [18, 19] as predicted by Lars Onsager in 1949 [20]. However, these experiments are generally limited to the regime of weak atom-atom interactions, where the Gross-Pitaevskii equation provides a microscopic model for the dynamics of the superfluid. Despite the progress in ultracold gases, where the regime of strong interactions can be reached by tuning the atomic scattering length, technical challenges have limited investigations of nonequilibrium phenomena [21, 22]. The strongly interacting regime defies a microscopic theoretical treatment and is the relevant regime for superfluid helium as well as for astrophysical superfluid phenomena such as pulsar glitches [23] and superfluidity of the quark-gluon plasma in the early universe [6]. The vortex dynamics in this regime are typically predicted using phenomenological vortex models. However, whether the vortices should have inertia [24, 25], the precise nature of the forces they experience due to the normal component of the fluid [26], and how to treat dissipation given the non-local nature of the vortex flow fields [27, 28] all remain unclear. Moreover, point-vortex modelling offers only a limited insight into the processes of vortex creation and annihilation, which are crucial for understanding of the dynamics of topological phase transitions.

This chapter reports the observation of coherent vortex dynamics in a strongly interacting two-dimensional superfluid. We achieve this by utilizing our microscale photonic platform, outlined in Chap. 2, to initialize vortex clusters in two-dimensional helium-4, confine them, and image their spatial distribution over time. Our experiments characterize vortex distributions via their interactions with resonant sound waves, leveraging ultraprecise sensing methods from cavity optomechanics [29–33]. Microscale confinement greatly enhances the vortex-sound interactions, and enables resolution of the dynamics of few-vortex clusters in a single-shot and tracked over many minutes, as they interact, dissipate energy and annihilate. We observe evaporative heating where the annihilation of low-energy vortices causes an increase in the kinetic energy of the remaining free vortices. A striking feature of our experiments is that vortex annihilation events draw energy out of a background flow, causing a net increase in free-vortex kinetic energy as the system evolves.

Our experiments yield a vortex diffusivity five orders-of-magnitude lower than has been observed previously for unpinned vortices in superfluid helium film [34]. This verifies that the diffusivity can become exceptionally small when operating at temperatures far below the superfluid transition temperature, as conjectured from extrapolation of thirty-year-old experimental observations [27, 34]. As a consequence, our system operates well within the regime of coherent vortex dynamics, with the timescale for dissipation found to exceed the coherent evolution time by more than five orders of magnitude. The on-chip platform reported here provides a new technology to explore the dynamics of phase transitions and quantum turbulence in strongly interacting superfluids, and to study how such fluids evolve towards thermal equilibrium and how they dissipate energy. It may also allow new phenomena to be engineered through strong sound-vortex interactions, as well as enabling the development of superfluid matter-wave circuitry on a silicon chip.

5.2 Sound-Vortex Interactions Enable Vortex Dynamics Imaging

In order to understand vortex dynamics in two-dimensional strongly interacting superfluids, it is crucial to be able to detect these dynamics experimentally. For instance, in bulk three-dimensional helium vortex dynamics have been observed with direct optical imaging techniques [35]. However, in two-dimensional helium the interaction between light and vortices is extremely weak due to the few-nanometer film thickness, Ångström scale of vortex cores, and the exceedingly low refractive index of superfluid helium. This precludes direct optical imaging techniques in two-dimensional helium, in contrast to the three-dimensional case. Instead, in our experiments the vortex dynamics are tracked via their influence on sound waves. Vortex-sound interactions are transduced via sound-light coupling and read out with ultraprecise methods of cavity optomechanics [31]. The vortices and sound waves coexist in a thin superfluid helium film. They are geometrically confined on the bottom surface of a microtoroidal cavity which supports optical whispering gallery modes (WGM) and is held above a silicon substrate on a pedestal. Figure 5.1 illustrates the interactions between vortices (Fig. 5.1a(top)) and sound modes (Fig. 5.1a(bottom)) confined to the surface of a disk. The sound modes are third-sound waves [8] elaborated in detail in Sect. 1.10 of this thesis. They are well described by resonant Bessel modes of the first kind, which are characterized by their radial m and azimuthal n mode numbers [36] (see Sect. 5.4.1 for details).

The vortex flow field causes Doppler shifts of the frequencies of the sound modes, lifting the degeneracy between clockwise (CW) and counter-clockwise (CCW) waves

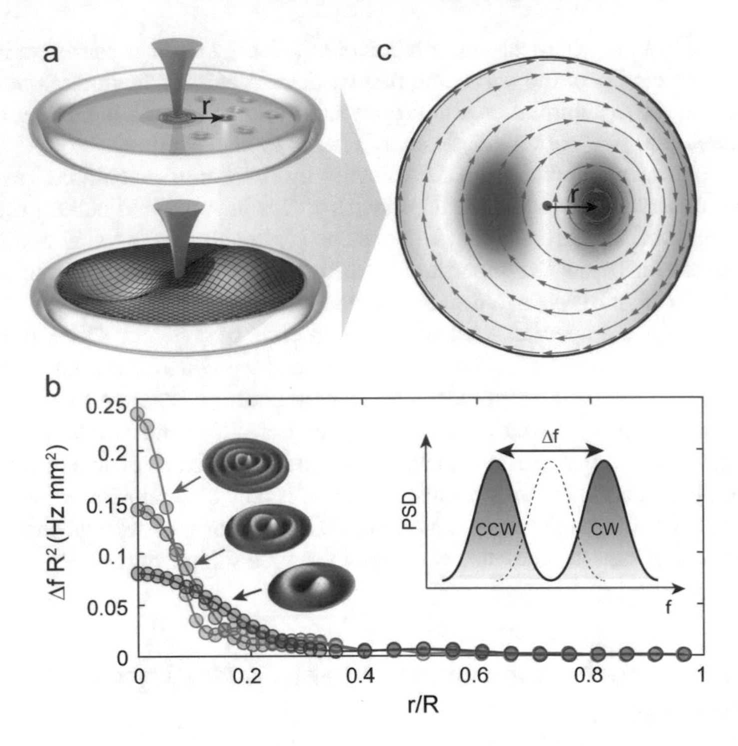

Fig. 5.1 **a** top: free vortices (blue) and pinned circulation (red) on the bottom surface of the microtoroid. bottom: a third-sound mode on the same surface. Confinement of both vortices and sound within the same microscale domain enhances both the vortex-sound interaction rate and the resulting frequency splitting between counterpropagating third-sound modes. Vortices and sound couple via their flow fields, as shown in (**c**). Here, the surface color represents the sound mode amplitude profile and the blue lines are streamlines of a single vortex offset from the disk origin by distance r. **b** Normalized frequency splitting per vortex for sound modes $(m, n) = (1, 3)$, $(1, 5)$ and $(1, 8)$ calculated via finite-element modelling [39], with their respective spatial profiles. Splitting is normalized such that it does not depend on size of the confining domain, i.e. $\Delta f \times R^2 \sim const$. The inset schematically depicts the vortex-induced splitting Δf between clockwise and counter-clockwise sound modes in the presence of a CW vortex

(Fig. 5.1b, c) in an effect known as Bryan's effect [37–39]. The magnitude of the frequency shift induced by a vortex depends both on its position and on the spatial profile of the sound mode, as shown for several modes in Fig. 5.1b. Moreover, the vortex-sound interaction strength is inversely proportional to the area of the domain confining vortices and sound waves. Thus, by confining vortices and sound to smaller areas, it is possible to greatly enhance the vortex-sound coupling and, hence, increase the resolution of vortex dynamics. The area of microscale confinement provided by the microtoroidal cavity in our experiments is four orders of magnitude smaller than in earlier works [38, 40], enabling a four order-of-magnitude enhancement of vortex-sound interactions, associated frequency splittings, and the vortex resolution (for detailed discussion of the vortex-sound coupling see Chap. 4).

Despite the strong interactions between helium atoms, the phase coherence and incompressibility of the superfluid combine to ensure linearity in our experiments. As such, the total flow field of a vortex cluster is given by the linear superposition of the flow of each constituent vortex. The total splitting between counter-rotating sound modes is then equal to the sum of the splittings generated by each vortex. We exploit this linearity, combined with the vortex-position dependent interaction and simultaneous measurements of splitting on several sound modes, to characterize the spatial distribution of vortex clusters in a manner analogous to experiments that use multiple cantilever eigenmodes to image the distribution of deposited nanoparticles [41].

5.3 Experimental and Data-Processing Details

5.3.1 Experimental Details

Our experimental system consists of a WGM microtoroidal optical cavity (Fig. 5.2a) placed inside a vacuum- and superfluid-tight sample chamber at the base of a closed-cycle ^{3}He cryostat (Fig. 5.2b). The sample chamber is filled with ^{4}He gas, with a pressure of $P = 70$ mTorr at temperature $T = 2.9$ K. Cooling down across the superfluid phase transition temperature condenses the ^{4}He gas into a superfluid film of thickness $d \sim 7.5$ nm, coating the inside of the chamber including the optical microcavity [31, 42]. For the ^{4}He gas pressure used in the experiment (70 mTorr) the superfluid phase transition temperature is around 1 K. The experiment is carried out at $T = 500$ mK.

The microcavity is fabricated from a wafer made of a 2 μm thick thermal oxide layer grown atop a silicon substrate (Fig. 5.2a). The high-quality thermally grown oxide ensures that both top and bottom surfaces of the fabricated resonator have typically <1 nm RMS roughness [43], leading to reduced vortex pinning (see Sect. 5.11). The silica microtoroid is elevated from the silicon substrate by a silicon pedestal (see Sect. 1.9). Light from a NKT Photonics (Koheras Adjustik) fibre laser with wavelength $\lambda = 1555.065$ nm is evanescently coupled into a resonator WGM (linewidth $\kappa/2\pi \approx 22$ MHz) via an optical nanofibre.

Motion of the superfluid film manifests as fluctuations of the phase of light confined inside the cavity, which are resolved via balanced homodyne detection (New Focus 1817 low-noise photodetector) implemented within a fibre interferometer (Fig. 5.2b) (see Sect. 2.4.2). The photocurrent is recorded with an Agilent Technologies MSO7104A oscilloscope with a sampling rate of 2 MHz. Frequency analysis of the output photocurrent reveals third-sound modes, with resonance frequencies in a good agreement with the expected Bessel mode frequencies (Fig. 5.3a) (for the details of modes identification see Sect. 5.4).

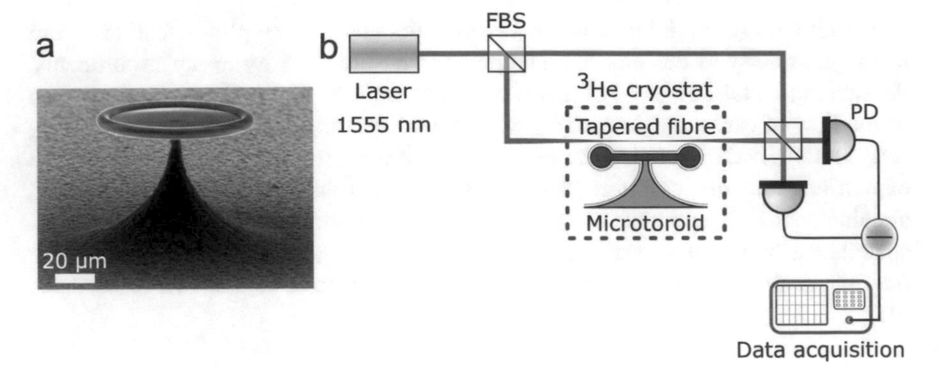

Fig. 5.2 **a** Scanning electron microscope image of the microtoroidal optical cavity used in the experiments. Scale bar: 20 μm. **b** Experimental set-up: balanced homodyne detection scheme implemented within a fibre interferometer. FBS: fibre beam splitter, PD: photodetector

5.3.2 Frequency Splitting Data Processing

In order to retrieve the frequency splitting evolution data shown in Fig. 5.3c, we record the balanced-detector photocurrent with an oscilloscope. The acquired data consists of 6 consecutive traces, separated by short gaps corresponding to the data saving time. Each trace is 50 s long and contains 100 million points (sampling rate 2 MHz). Each 50 s trace is then broken down into 125 bins, each 0.4 s long. We Fourier-transform each bin in order to obtain $6 \times 125 = 750$ third-sound power spectra, as shown in Fig. 5.3a. We fit each tracked sound mode with a double-peaked Lorentzian function in order to acquire the frequency separation between the split peaks, see Fig. 5.3c. This allows us to plot the frequency splitting for the five tracked sound modes as a function of time, with 750 time steps during the 360 s experimental decay process.

5.3.3 Geometric Contribution to Sound Splitting

Small deviations from circular symmetry of the resonator, for instance formed during the microtoroid fabrication process, may split degenerate third-sound modes even in the absence of vortices in the superfluid film. This geometric splitting [39, 40] of the sound modes is analogous to the roughness-induced optical doublet splitting well known in high-Q optical microcavities [44].

In our experiment, most sound modes exhibit some small ($<$1 kHz) degree of geometric splitting. This splitting is easily distinguishable from the vortex-induced splitting, as it is of smaller magnitude, constant in time, and can be determined by observing the native splitting present when repeatedly cycling through the superfluid transition temperature. To account for it, we remove its contribution to the total splitting Δf_{total} in order to isolate the vortex-induced contribution Δf_{vortex} using the

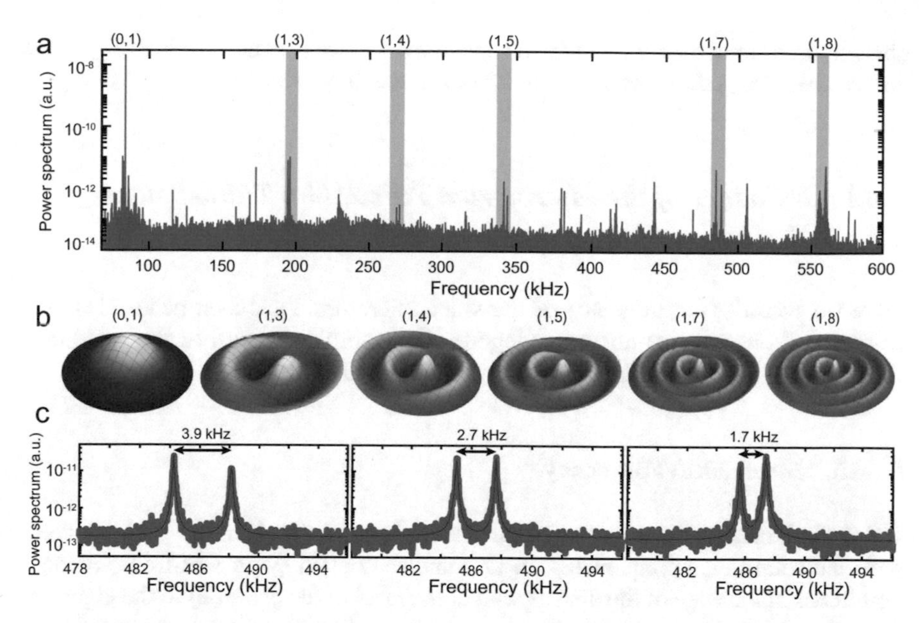

Fig. 5.3 **a** Third-sound power spectrum. The five sound modes monitored in the experiments are highlighted in red. The red bars are positioned at the theoretically predicted sound eigenfrequencies of a 30 μm radius disk resonator with free boundary conditions and a 7.5 nm film thickness [36], in good agreement with the experimental frequencies. Note that mode (1,6) is not used in the experiments, as its proximity to another sound mode precludes reliable peak fitting. **b** Surface deformation profiles of the 6 sound modes labelled in (**a**). **c** Example of the mode splitting decay observed in the experiments. The three panels show, from left to right, the splitting decay of mode (1,7) as the experiment progresses. Red line: double Lorentzian peak-fit to the data

following relationship [40]:

$$\Delta f_{\text{vortex}} = \sqrt{\Delta f_{\text{total}}^2 - \Delta f_{\text{geo}}^2} \tag{5.1}$$

Data presented from here onwards corresponds to the vortex-induced splitting solely, with the geometric splitting contribution removed.

5.4 Third-Sound Modes Identification

Figure 5.3a shows a power spectrum of light transmitted through the optical microcavity, revealing the presence of multiple superfluid third-sound modes. Correct identification of the sound modes is important in order to ascertain how each sound mode is differently affected by the presence of vortices [39]. Since sound modes are confined to the surface of the microtoroidal cavity which rests upon the pedestal, then, in order to correctly identify the sound modes, we first need to investigate the influence of the pedestal on frequencies of the sound modes.

5.4.1 *Influence of the Microtoroid Pedestal on Third-Sound Modes*

Here we consider the influence of the silica microtoroid's silicon pedestal on the third-sound eigenmodes on the underside of the toroid, as well as the pedestal's influence on the splitting experienced by these sound waves.

5.4.1.1 Mechanical Frequency

Using the Finite-Element Method (FEM) techniques outlined in Ref. [39], we compare the sound eigenfrequencies on an annular domain (with free-free boundary conditions at the edge of the disk as well at the level of the pedestal) to the eigenfrequencies on a simple disk geometry of same outer radius. The results are summarized in Table 5.1. The effect of the pedestal on the third-sound frequencies is negligible, typically less than 1%. This is because the presence of the pedestal does not significantly alter the sound eigenmode shapes, as shown for three different eigenmodes in Fig. 5.4 and discussed below.

5.4.1.2 Sound Modes Frequency Splitting

Similarly, we compare the sound mode frequency splitting Δf due to a vortex at the center of a disk geometry to the splitting due to a persistent current around the pedestal in an annular domain. The results are also summarized in Table 5.1. These values are computed using the FEM simulation techniques described in Ref. [39]. The splitting differences are again small, typically below 3%. For this reason, it is reasonable to neglect the presence of the pedestal and model the underside of the microtoroid as a circular disk resonator, so far as the mechanical eigenmodes and the splitting they experience are concerned. (Note that the presence of the pedestal is taken into account in the kinetic energy calculations discussed in Sect. 5.5).

5.4.1.3 Third-Sound Eigenmodes on a Disk and Annulus

Here we briefly describe the analytical expressions of the superfluid third-sound eigenmodes on a disk and on an annulus. The eigenmodes on a disk are given by [36]:

$$\eta_{m,n}(r,\theta) = \eta_0 \, J_m\left(\xi_{m,n}\frac{r}{R}\right)\cos(m\theta)\,, \tag{5.2}$$

Table 5.1 Comparison table for the eigenfrequencies and splittings on a disk versus annular geometry. Frequency and splitting values quoted for the eigenmodes of a 30 μm radius disk/a 30 μm outer radius annulus (see eigenmodes in Fig. 5.4). Frequency values are quoted for a 10 nm thick superfluid film. The frequency error (f_{error}) and splitting error (Δf_{error}) are typically quite small, below 3%. Modes with $m = 0$ cannot be decomposed in the basis of CW and CCW rotating modes and experience therefore no splitting

Mode	f_{disk} (Hz)	f_{annulus} (Hz)	Δf_{disk} (Hz)	$\Delta f_{\text{annulus}}$ (Hz)	f_{error} (%)	Δf_{error} (%)
(0,1)	57318	57426	–	–	0.19	–
(1,6)	269484	267301	193.3	188.3	0.81	2.6
(1,8)	363633	360127	261.8	256.7	0.96	1.95

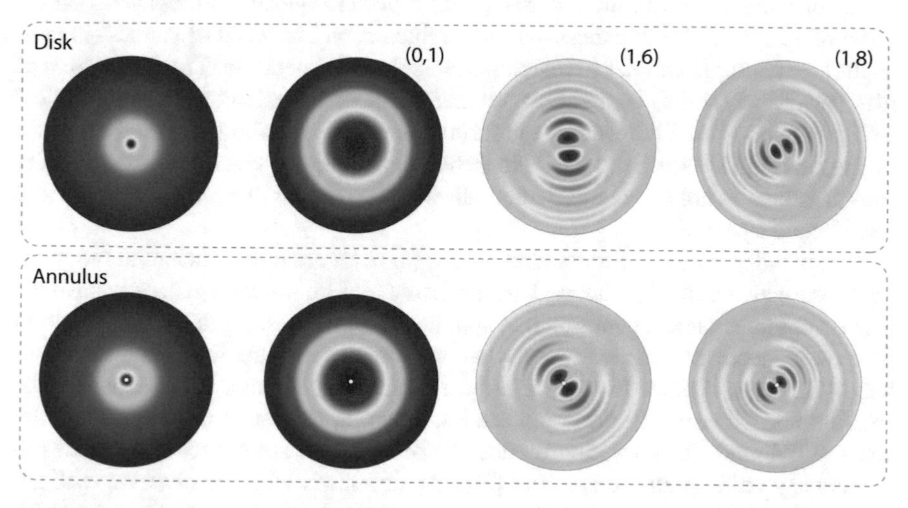

Fig. 5.4 Leftmost image for both disk and annulus: calculated flow field due to quantized circulation around a centered point-vortex (disk) or a persistent current around a pedestal with a 1.5 μm diameter (annulus). Rightmost images: Out-of-plane displacement profiles for three distinct sound eigenmodes. Color code: red—positive, blue—negative, green—no displacement. The (m,n) numbers respectively correspond to the mode's azimuthal and radial orders, see Eqs. (5.2) and (5.3)

where $\eta_{m,n}$ describes the out-of-plane deformation of the superfluid surface for the (m, n) mode as a function of polar coordinates r and θ. The (m,n) numbers respectively correspond to the mode's azimuthal and radial orders, and $\xi_{m,n}$ is a frequency parameter depending on the mode order and the boundary conditions. The eigenmodes on an annulus [36] are given by:

$$\eta_{m,n}(r,\theta) = \eta_0 \left(J_m\left(\xi_{m,n}\frac{r}{R}\right) + \alpha Y_m\left(\xi_{m,n}\frac{r}{R}\right)\right)\cos(m\theta), \tag{5.3}$$

where J_m and Y_m are respectively Bessel functions of the first and second kind of order m. For small values $r_p \ll R$, as is the case in our experiments, $\alpha \ll 1$ as Y_m

diverges at $r = 0$, and the eigenmodes of the annulus are very close to those of the disk.

5.4.2 Modes Identification

As we discussed above, the presence of the pedestal has a negligible effect on the sound eigenfrequencies, and, thus, the experimental spectrum can be fitted with the eigenfrequencies of a disk. We find that our experimental spectrum can be well reproduced by the eigenfrequencies of a 30 μm radius disk resonator with free boundary conditions and a 7.5 nm film thickness, with typical frequency discrepancies on the order of 1%. The film thickness—via its influence on the speed of sound—is used as the sole fitting parameter to match the Bessel mode spectrum. The fitted value of film thickness (7.5 nm) is in good agreement with values extracted from the optical cavity frequency shift induced by the formation of the superfluid film on the surface of the cavity [31] (see Sect. 2.4). Note that the frequency spacing of higher-order Bessel modes is not harmonic, which allows us to discriminate between free and fixed boundary conditions.

Rotationally invariant third-sound modes ($m = 0$) are most efficient at modulating the effective optical path length of the resonator [36, 45], and generally have the highest optomechanical coupling rate and, hence, the largest signal-to-noise ratio in the power spectrum. This is indeed what we observe, with the fundamental 'drumhead' (0,1) Bessel mode having the largest signal-to-noise ratio in the experiments, see Fig. 5.3a. Rotationally invariant modes, however, cannot be decomposed in the basis of CW and CCW rotating modes. As such, we do not expect these modes to display any vortex-induced splitting [39]. Again, this is what we observe, with the (0,1) mode remaining unperturbed by the vortex generation process, while the (1,3), (1,4), (1,5), (1,7) and (1,8) modes all exhibit splitting, as seen in Fig. 5.3a.

5.5 Influence of the Microtoroid Pedestal on Vortex Dynamics

Even though the influence of the microtoroid pedestal on third-sound modes' frequencies and splittings is negligible, we find that the presence of the pedestal introduces significant qualitative changes to vortex dynamics. Namely, the pedestal creates a deep potential that vortices can pin to. Thus, we shall now elaborate on how the pedestal modifies the physics of our system. We do so by considering the kinetic energy landscape of the fluid confined within the circular boundary of the microtoroid.

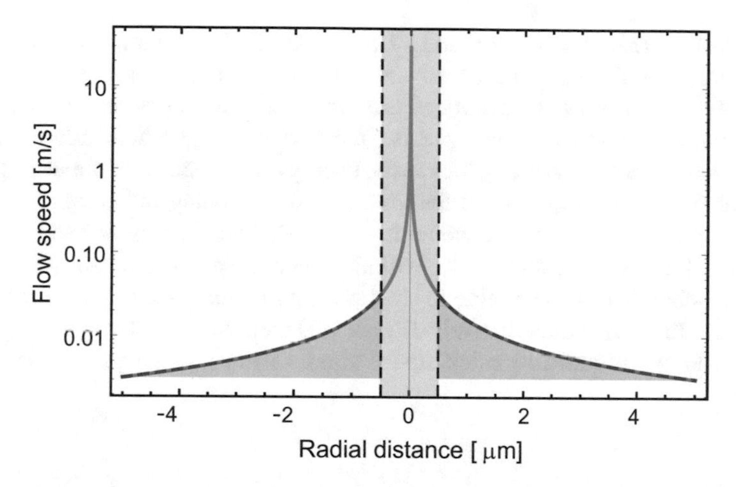

Fig. 5.5 Superfluid flow velocity as a function of the radial distance from the vortex core, for a single circulation quantum $\kappa = h/m_{\text{He}} = 9.98 \times 10^{-8}\ \text{m}^{-2}\text{s}^{-1}$ (blue line). Clipping the high-velocity component of the flow field makes quantized circulation around the μm-sized microtoroid pedestal (depicted by the greyed-out area) energetically favourable

5.5.1 *Pinning Potential of the Pedestal*

While superfluid flow is purely irrotational, superfluids may contain vorticity consisting of irrotational flow around a topological defect in the fluid [46, 47]. This topological defect may take the form of an Ångström-sized normal-fluid core in ^{4}He, or a macroscopic engineered barrier in a non-simply connected geometry. The kinetic energy E_k associated with such a flow around a centred circular defect in a circular domain of radius R in two-dimensional superfluids is given by:

$$E_{k,n} = \frac{1}{2}\rho_s \frac{(N\kappa)^2}{2\pi} \ln\left(\frac{R}{a_0}\right), \tag{5.4}$$

with ρ_s being the superfluid surface density, a_0 the defect radius and $N \times \kappa$ the quantized circulation around the defect. Here, $\kappa = h/m_{\text{He}} = 9.98 \times 10^{-8}\,\text{m}^{-2}\text{s}^{-1}$ is a single circulation quantum in ^{4}He, with m_{He} being the mass of a helium atom and h Planck's constant. Given that circulation around a macroscopic barrier—such as the microtoroid pedestal ($a_0 = r_p \sim 10^{-6}$ m)—clips the high-velocity region of the flow (see Fig. 5.5), it is energetically favourable compared to circulation around a vortex core in the film ($a_0 \sim 10^{-10}$ m). This has been observed via stirring of superfluids in annular containers, where the vorticity preferentially takes the form of persistent flow around the inner annular boundary, and vortices appear in the fluid itself only for much larger rotation speeds [48, 49].

If we consider a normal-fluid-core vortex centred on a disk of radius $R = 30\,\mu\text{m}$, then the kinetic energy of its flow is calculated to be $\sim 1 \times 10^{-20}$ J, which corresponds to $\sim$2000 $k_B T$ (for $T = 500$ mK and a 10 nm thick film). In contrast, for our $R = 30$ µm resonator with pedestal radius $r_p \sim 1$ µm there is approximately *three* times less energy in the flow for quantized circulation around the pedestal than around the normal-fluid core. However, placing increasingly large circulation around the pedestal becomes less advantageous as the energy cost of each additional circulation quantum $E_{k,N+1} - E_{k,N}$ is amplified by $2N$, with N being the number of quanta already present in the fluid, as shown in Eq. (5.4). This thereby provides an upper bound on the possible number of circulation quanta around the pedestal before it becomes energetically favourable to shed circulation quanta and form free vortices in the fluid. In order to quantitatively estimate the metastability of the pinning around the pedestal, we numerically calculate the kinetic energy barrier for the shedding of vortices.

5.5.2 Calculation of the Flow Kinetic Energy

There are two mechanisms through which the large quantized circulation pinned to the pedestal may decay. First, through the escape of vortices of the same sign as the circulation (without loss of generality, for convenience, here we assume the pinned circulation to be 'positive' and the free vortices to be 'negative'). Second, a negative vortex may be generated elsewhere and migrate inwards, annihilating circulation on the pedestal. We estimate the change in kinetic energy of the flow for these two processes. In order to do so, we first calculate the flow field, exploiting streamfunctions for both free vortices and the circulation pinned to the pedestal.

The streamfunction Ψ_{free} of a free point vortex on the underside of the toroid can be approximated in cartesian coordinates by [39, 50–52]:

$$\Psi_{\text{free}} \simeq \frac{\kappa}{2\pi}\left(\ln\left(\sqrt{(x-X_1)^2+y^2}\right) - \ln\left(\sqrt{(x-X_2)^2+y^2}\right)\right.$$
$$\left. + \ln\left(\sqrt{x^2+y^2}\right) - \ln\left(\sqrt{(x-X_3)^2+y^2}\right)\right). \tag{5.5}$$

Here X_1 is the radial coordinate of the vortex (along the x axis), $X_2 = \frac{r_p^2}{X_1}$ is the radial coordinate of the opposite-circulation image vortex required to enforce no-flow across the pedestal boundary, and $X_3 = \frac{R^2}{X_1}$ is the radial coordinate of the opposite-circulation image vortex required to enforce no-flow across the resonator boundary, as shown in Fig. 5.6b. Since there are two boundaries in the problem, the image vortices also require their own image vortices, leading to an infinite series. However, for small values of $r_p \ll R$, as is the case in our experiments, the infinite series can safely be truncated after the first term, as in Eq. (5.5), while conserving the absence of flow through the resonator and pedestal boundaries, as visible in the calculated streamlines in Fig. 5.6a, b.

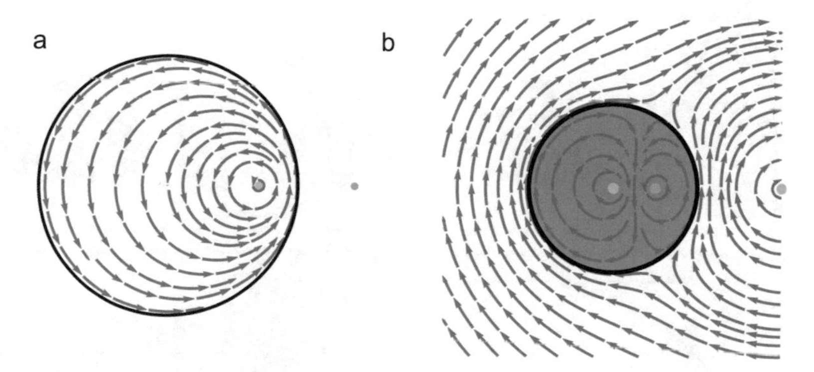

Fig. 5.6 **a** Flow streamlines of a point vortex (red dot) offset from the center inside a circular domain (black circle). The no-flow boundary condition is enforced by the presence of an opposite-sign image vortex outside the boundary (blue dot) [50]. **b** Flow streamlines of a point vortex (rightmost red dot) outside a cylindrical obstacle (grey disk) in a two-dimensional plane. The no-flow boundary condition is enforced by an opposite-sign image vortex inside the obstacle (blue dot) and a same-sign image vortex in the center of the obstacle (left red dot)

The streamfunction Ψ_{pedestal} of the quantized circulation around the pedestal is given by:

$$\Psi_{\text{pedestal}} = \frac{N\kappa}{2\pi} \ln\left(\sqrt{x^2 + y^2}\right), \tag{5.6}$$

where N is the number of circulation quanta pinned around the pedestal.

The total streamfunction Ψ_{tot}, describing the combined flow, is then simply $\Psi_{\text{tot}} = \Psi_{\text{free}} + \Psi_{\text{pedestal}}$. From the streamfunction Ψ_{tot}, the superfluid velocity components are given by

$$v_x = \frac{\partial \Psi_{\text{total}}}{\partial y} \quad \text{and} \quad v_y = -\frac{\partial \Psi_{\text{total}}}{\partial x}. \tag{5.7}$$

The kinetic energy of a given flow field is obtained through numerical integration of the calculated field (Eq. (5.7)) over the annular region contained within $r_p < r < R$.

5.5.3 *Metastability of the Flow Pinned Around the Pedestal*

Figure 5.7a, b illustrate two alternate decay scenarios for a large circulation trapped around the microtoroid pedestal. First, a positive vortex (red) can be shed from the large positive circulation pinned around the pedestal. Second, a negative incoming vortex (blue) can annihilate with the positive circulation on the microtoroid pedestal. The energy cost of both these scenarios is illustrated in Fig. 5.7c. The red curve shows the energy difference between the two following configurations:

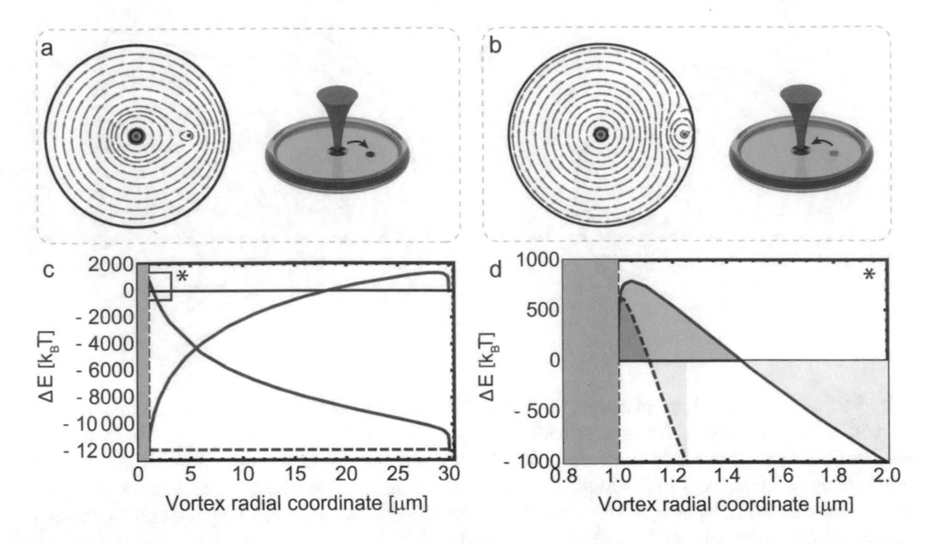

Fig. 5.7 Two alternate decay scenarios for a large ($\kappa \gg 1$) circulation trapped around the micro-toroid pedestal. **a** Positive vortex (red) 'peeling off' from the large positive pinned circulation around the central pedestal (greyed-out region). Blue arrows display the flow streamlines calculated from Eq. (5.7). **b** Incoming negative vortex (blue) annihilating with the positive circulation on the micro-toroid pedestal. **c** Calculated kinetic energy cost of the decay processes illustrated in (**a**) and (**b**), starting from a positive circulation of $11 \times \kappa$ around the pedestal (red curve—escaping positive vortex, blue curve—incoming negative vortex). The dashed red line marks the theoretical energy drop from the removal of one circulation quantum, i.e. the drop from $11 \times \kappa$ to $10 \times \kappa$. **d** Zoom-in of the region marked by an asterisk '*' in (**c**), highlighting the potential barrier for the shedding of positive vortices. Solid red line and dashed red line show the potential barriers starting from a positive circulation of $11 \times \kappa$ and $31 \times \kappa$ around the pedestal respectively. Greyed-out region marks the location of the pedestal

i. a positive circulation of 10κ on the pedestal and a free point vortex of circulation κ at radial coordinate x
ii. a positive circulation of 11κ on the pedestal

If the free vortex can move far enough away from the pedestal, this escape process is energetically favourable ($\Delta E < 0$), as the flow fields of the free vortex and the macroscopic circulation no longer sum up constructively. However, as shown in Fig. 5.7d, there is an initial energy barrier on the order of several hundred $k_B T$ for the free vortex to break free from the topological defect. Indeed, in the initial stage of the vortex shedding process the flow fields of the pinned circulation and the free vortex still add up constructively, and do not offset the additional energy cost of the high velocity flow near the normal-fluid core, as shown in Fig. 5.5.

As discussed previously, the overall shape of the energy barrier is dependent on the number of pinned circulation quanta N. For increasing circulation around the pedestal the barrier becomes narrower and shallower (dashed red line in Fig. 5.7d), but the conclusion remains identical. While the lowest energy state of the system corresponds to no persistent current, a large energy barrier (on the order of several hundred $k_B T$ for the vortex numbers present in the experiment) explains why the circulation remains pinned around the microtoroid pedestal for the duration of the experiment.

Similarly, the creation of a negative vortex on the microtoroid boundary also incurs a large energy penalty due to the energy cost of the high-velocity region near the vortex core (see Fig. 5.7c). However, for radial coordinates $r < 18\ \mu\text{m}$, that energy cost is larger than the offset by the cancellation of the pinned persistent current flow field, so that the negative free point vortex introduces 'negative' kinetic energy in the system.

5.6 Generation of Nonequilibrium Vortex Clusters

Having described technical details of our experiments and provided theoretical estimations of how the presence of the microtoroid pedestal should modify the dynamics of our vortex-sound system, we are now in position to move on to the experimental results and their discussion. We start off by describing the generation of nonequilibrium vortex clusters with laser light.

We find experimentally that vortex clusters can be optically initialized on the surface of the microtoroid in several ways, including pulsing the intensity of the injected laser to induce superfluid flow via the '*fountain effect*' [46], and optomechanical driving of low-frequency third-sound modes via dynamical backaction [31]. Both of these techniques induce flow that exceeds the superfluid critical velocity [46] at the interface between the pedestal and the microtoroid, triggering the generation of vortex pairs (Fig. 5.8a(top)). In the presence of a circular boundary, an ensemble of vortex pairs evolves into a metastable state characterized at high energies by a large-scale negative-temperature Onsager vortex dipole [18, 20]. As we discussed earlier, the microtoroid pedestal creates a deep potential that generated vortices can get trapped onto (see Sect. 5.5). Vortices of one sign become pinned, creating a macroscopic circulation around the pedestal, while vortices of the opposite sign evolve into a free-orbiting metastable cluster, as shown in Fig. 5.8a(bottom) (for details see Sect. 5.9.2). Although this metastable state would be characterised as a negative-temperature vortex dipole even in the absence of the pinning site, in our case the presence of the microtoroid pedestal with its pinning potential introduces qualitatively different features into the the dynamics of the metastable state.

The timescale within which a metastable state is reached scales inversely with the vortex-vortex interaction strength and hence depends on the density of free vortices. This timescale can be estimated from the characteristic turnover time for internal rearrangement of the free-vortex cluster, $\tau \sim r_c^2/N\kappa$, where N is the number of free vortices, r_c is the radius of the cluster, and κ is the circulation quantum [53]. Taking the case of two free vortices separated by the disk radius $R = 30\ \mu\text{m}$ provides an upper bound to the turnover time of $\tau \lesssim 5$ ms.[2] This is substantially faster than

[2] As we will see later on, in Sect. 5.9.2, this timescale is also confirmed via point-vortex simulations.

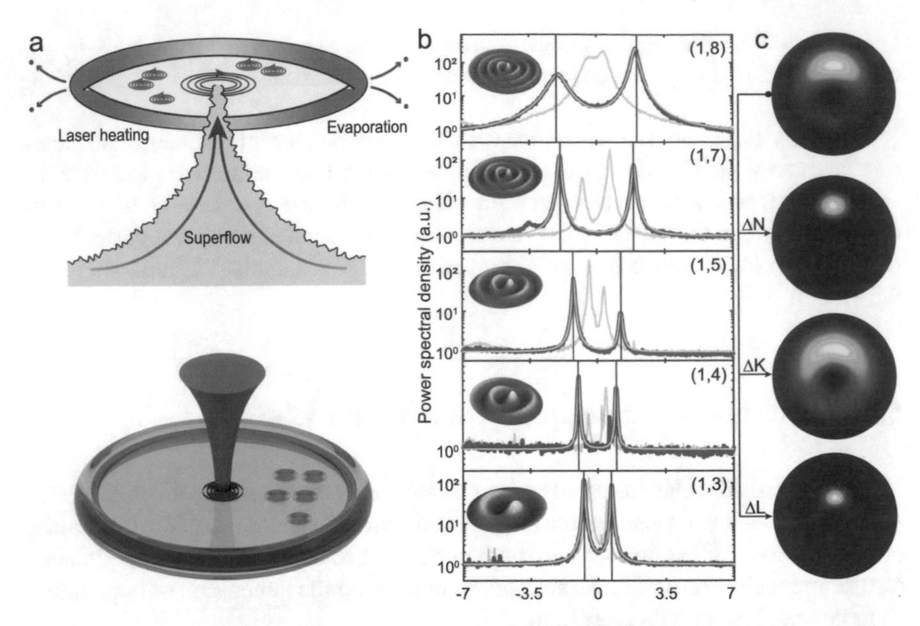

Fig. 5.8 Laser initialization and observation of vortex clusters. **a** Top: sketch of the vortex generation process. Optically induced superfluid flow exceeds the superfluid critical velocity, resulting in the generation of vortex pairs. An ensemble of vortex pairs evolves into a metastable vortex dipole. Bottom: artist's representation of the metastable vortex state. **b** Splittings of $(m, n) = (1, 3), (1, 4), (1, 5), (1, 7)$ and $(1, 8)$ third-sound modes immediately after the initialization process. Spatial amplitude profiles of the split modes are shown as insets. Red bars mark the theoretically computed splittings right after the system initialization. Grey spectra correspond to the unperturbed third-sound modes without vortex ensemble initialization. The residual splitting in these unperturbed spectra is due to irregularities in the circularity of the microtoroid that break the degeneracy between standing-wave Bessel modes. We accounted for these in data processing (see Sect. 5.3.3). **c** Exemplar simulated metastable distributions showing the effects of vortex annihilation and of changes in total kinetic energy K and angular momentum L. The colormap indicates the free-vortex probability density. Top metastable state: $\{N, K, L\} = \{10, 0.43\,\text{aJ}, 120\,\text{ag}\,\mu\text{m}^2\,\text{s}^{-1}\}$; second-to-top: $N \to 9$; second-to-bottom: $K \to 0.41\,\text{aJ}$; bottom: $L \to 44\,\text{ag}\,\mu\text{m}^2\,\text{s}^{-1}$

both the dissipation of the system and the temporal resolution of our measurements. Consequently, the vortex cluster can be well approximated to exist in a metastable state throughout its evolution, with this state modified continuously by dissipation and in discrete steps by vortex annihilation events.

Each possible metastable state is uniquely characterized by the number of free vortices, kinetic energy and angular momentum. Performing point-vortex simulations we determine the possible metastable vortex distributions as a function of these three parameters (see Sect. 5.9.2). Next, in order to determine the instantaneous metastable vortex distribution at each time of the cluster evolution, we use the vortex-position dependent splitting function $\Delta f(r)$ (see Fig. 5.1b) to compute frequency shifts expected on the $(m, n) = (1, 3), (1, 4), (1, 5), (1, 7), (1, 8)$ third-sound modes for each possible metastable distribution and compare these to the experimentally observed shifts. This allows us to ascertain both the metastable state that most closely matches the observed frequency shifts at a given time, as well as the range of metastable states for which the shifts are statistically indistinguishable.

Figure 5.8b shows the experimentally observed frequency shifts at the start of the measurement run, just after the vortex cluster has been initialised. As shown by the vertical red bars, we find an excellent agreement between the observed frequency splittings and those computed with the optimal metastable distribution. Moreover, although the measurements only loosely constrain the angular momentum (see Sect. 5.8), we find that the statistical uncertainty in the free-vortex number and kinetic energy is relatively small. We identify the initial metastable vortex distribution as a vortex dipole either containing 17 free vortices with a total kinetic energy of $K = 0.8^{+0.1}_{-0.03}$ aJ and angular momentum of $L = 0.3^{+0.05}_{-0.2}$ fg μm^2 s^{-1}, or 16 vortices with $K = 0.8^{+0.2}_{-0.02}$ aJ and angular momentum of $L = 0.2^{+0.3}_{-0.07}$ fg μm^2 s^{-1}. We are also able to determine the dipole separation, which in the case of 17 vortices is found to be $7.3^{+1.2}_{-0.8}$ μm (for details see Sect. 5.7).

In their metastable state, the free vortices exist in an orbiting horse-shoe shaped cluster separated from the origin, as illustrated in Fig. 5.8c. Together with the macroscopic circulation this forms a vortex dipole. Figure 5.8c illustrates how the distribution of the cluster evolves due to changes in vortex number, kinetic energy and angular momentum.

5.7 Vortex Decay Models

Once the vortex cluster is initialised, we simultaneously measure the frequency shifts induced on several third-sound modes as the cluster evolves over time. Continuously monitoring the superfluid sound modes reveals that their splitting decays over a timescale on the order of minutes (Fig. 5.9). The splitting decay is a direct consequence of the vortex cluster evolution. We developed a number of feasible vortex decay models which we compare here in order to ascertain which of these models characterises the splitting evolution most consistently over the entire course of the decay process.

Since the frequency splitting experienced by each third-sound mode due to the presence of a vortex is unique (see Fig. 5.1b and Ref. [39]), the ability to monitor the splitting on multiple sound modes simultaneously allows us to infer spatial information about the experimental vortex distribution. And it is this capability that we use to discriminate between several different vortex decay scenarios (Fig. 5.10a).

- Scenario 1.[3] The experimentally observed splitting is due to a tight cluster of same-signed vortices located at the centre of the microtoroid. Such a distribution could be initialized by a circular stirring of the superfluid film, resulting in a centripetal Magnus force strong enough to push vortices of one sign into a tight cluster at

[3]Note that Scenarios 1 and 2 assume vortex dynamics on the top surface of the microtoroid.

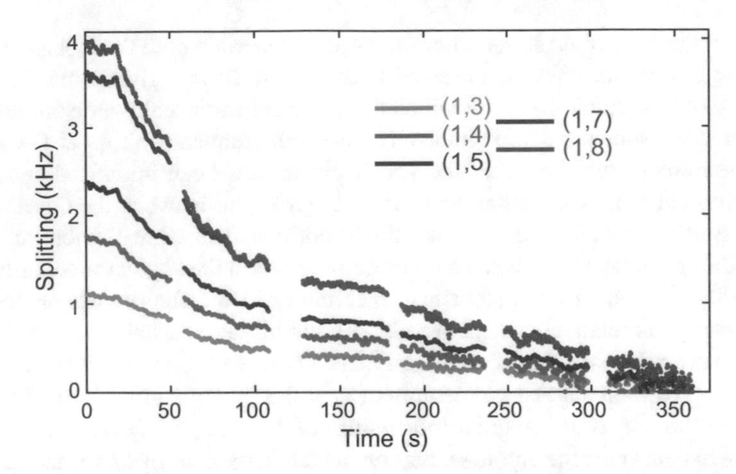

Fig. 5.9 Experimental splitting decay of the $(m, n) = (1, 8)$, $(1, 7)$, $(1, 5)$, $(1, 4)$ and $(1, 3)$ third-sound modes (top-to-bottom traces respectively). The raw data was recorded on a high bandwidth, high memory-depth oscilloscope. In this run, six continuous measurements were taken, separated by data-saving periods each of approximately ten seconds. Note that the data is displayed with no geometric contribution to the splitting (see Sect. 5.3.3)

the microtoroid origin, and vortices of opposite sign towards the resonator outer boundary where they either 'annihilate' with their images or disappear in any other way. After initialization, dissipation in the system will drive a radial expansion of the tightly-packed cluster. As discussed in Sect. 5.9.4, point-vortex simulations reveal that a tight vortex cluster will rapidly relax into an expanding flat-top (i.e. spatially uniform) distribution, irrespective of the initial cluster arrangement. This flat-top expansion scenario is depicted on the blue resonator in Fig. 5.10a.

- Scenario 2. The experimentally observed splitting is due to a tight cluster of same-signed vortices located at the centre of the resonator, as in scenario 1. The difference here lies in the presence of pinning sites (black stars in Fig. 5.10a) on the resonator surface. Expansion of the tightly-packed cluster in the presence of pinning sites resembles diffusive Brownian motion whereby vortices hop from pinning site to pinning site with a slow outward drift due to dissipation (see individual vortex trajectory marked by black arrows). Under such conditions the spatial probability density distribution of the expanding cluster can be approximated by a Gaussian (see Sect. 5.9.4). This Gaussian expansion scenario is depicted on the green resonator in Fig. 5.10a.
- Scenario 3. The experimentally observed splitting is due to a macroscopic persistent current around the pedestal at the centre of the resonator (see discussion of the pedestal pinning potential in Sect. 5.5.1). This persistent current then decays through the slow shedding of same-sign vortices. Due to dissipation, these vortices will rapidly spiral out of the resonator, such that the probability density can be modelled by a temporally decaying delta function in the centre of the resonator. Due to the strong potential energy barrier for the vortex shedding process (see

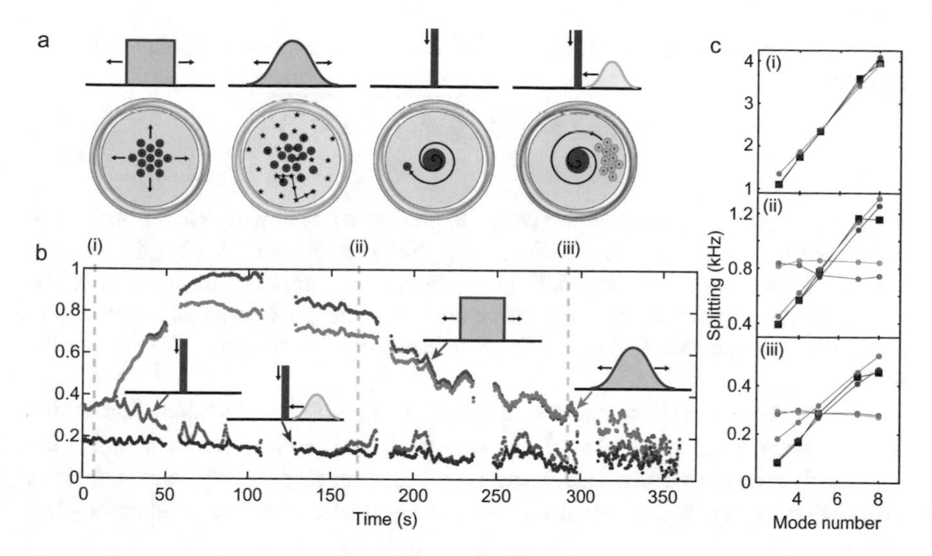

Fig. 5.10 **a** Four different vortex decay scenarios. From left to right: flat-top (uniform density) expansion of a vortex cluster (blue disk); Gaussian expansion of a vortex cluster with pinning sites (green disk); decay of a macroscopic circulation through shedding of same-sign vortices (orange disk); decay of a macroscopic vortex dipole consisting of a pinned macroscopic circulation (red) and a cluster of free opposite-sign vortices (purple disk). The associated probability density is plotted above each scenario. **b** Calculated RMS error for each decay scenario across the 360 s experimental duration. The trace colours keep the same colour scheme as in (**a**). At longer times, all scenarios converge towards a similar error as the experimental splitting has almost entirely decayed and our approach loses its discriminating power. **c** Experimentally-measured splitting of the five tracked sound modes ((1,3), (1,4), (1,5), (1,7) and (1,8)) (black squares) and simulated splitting corresponding to the four decay scenarios listed above. Each scenario is identified by the same colour code as in (**a**). Each sound mode is identified by its azimuthal mode number m, e.g. '8' corresponds to mode (1,8). Plots *i*, *ii* and *iii* correspond to the times indicated by the dashed grey lines in (**b**)

Sect. 5.5.3), such a scenario is unlikely but is nonetheless considered here for completeness. This persistent current decay scenario is depicted on the orange resonator in Fig. 5.10a.

- Scenario 4. The experimentally observed splitting is due to a macroscopic persistent current pinned to the pedestal at the centre of the microtoroid, and a cluster of orbiting free vortices of opposite sign. Such a distribution can be initialized by superfluid flow exceeding the critical velocity up the device pedestal, leading to the generation of pairs of opposite-sign vortices. One sign pins to the pedestal while the other forms into a freely orbiting cluster. After initialization, dissipation will cause the orbiting cluster to spiral inwards, where it will annihilate with the persistent current on the pedestal. This vortex dipole decay scenario is depicted on the purple resonator in Fig. 5.10a.

Each decay model is tested by comparing its predicted splitting on all sound modes to the experimental data. The figure of merit for this comparison is the root-mean-square (RMS) splitting error relative to the experimental data which is given by

$$\Delta f_{\text{RMS error}} = \sqrt{\sum_i \left(\Delta f_{\text{pr,i}} - \Delta f_{\text{exp,i}}\right)^2}. \tag{5.8}$$

Here, $\Delta f_{\text{pr,i}}$ and $\Delta f_{\text{exp,i}}$ are respectively the predicted and experimental third-sound mode splitting for mode i. Summation in Eq. (5.8) is performed over the five tracked sound modes identified in Fig. 5.3. The results of the comparison of each of the models to the experiment over the entire course of the decay process are shown in Fig. 5.10b. We now briefly describe how the RMS error is computed for each of the models.

Each of the models is characterized by its spatial probability density distribution $\rho_{\text{R}_{\text{dist}}}(r)$, normalised such that $\int_0^R \rho_{\text{R}_{\text{dist}}}(r)\, r\, \text{d}r = 1$. Decay scenarios 1 and 2 are characterised by the uniform and Gaussian distributions respectively. For the uniform distribution $\rho_{\text{R}_{\text{dist}}}(r)$ is constant within R_{dist} and zero elsewhere, i.e. it takes the form of

$$\rho_{\text{R}_{\text{dist}}}(r) = \begin{cases} \dfrac{2r}{\text{R}_{\text{dist}}^2} & \text{if } r < \text{R}_{\text{dist}}, \\ 0 & \text{if } r > \text{R}_{\text{dist}}, \end{cases} \tag{5.9}$$

which derives from the normalization condition. For the Gaussian distribution we proceed in the similar fashion. The splitting $\Delta f_{\text{pr,i}}^{[1,2]}$ predicted by each of these expansion models on the ith third-sound mode can then be obtained as

$$\Delta f_{\text{pr,i}}^{[1,2]} = N \int_0^R \rho_{\text{R}_{\text{dist}}}(r)\, \Delta f_{\text{th,i}}(r)\, r\, \text{d}r, \tag{5.10}$$

where $\Delta f_{\text{th}}(r)$ refers to the theoretically calculated vortex-induced splitting functions, as shown in Fig. 5.1b. These splitting functions are independently obtained for all five experimentally tracked sound modes through Finite-Element Method calculation of the overlap between the vortex and sound flow fields [39].

Both the flat-top and Gaussian distributions are uniquely characterized by an initial vortex number and the distribution radius. Looking at the initial experimental splitting ($t = 0$) of the five tracked sound modes and computing $\Delta f_{\text{pr,i}}^{[1,2]}$ for each of them, we find the combination of the vortex number ($N = 15$) and the distribution radius which most closely matches the experimental results ($\Delta f_{\text{RMS error}}$ is minimized), and the associated confidence intervals, as illustrated in Fig. 5.11. This initial vortex number is then kept constant throughout the fitting procedure, as the distribution expands with no annihilation events. At all subsequent experimental time steps, the optimal radius which minimizes the RMS error is inferred. The fitting procedure thus produces an estimate of the distribution radius versus time, as well as the RMS error of the fitting procedure versus time, as plotted in the blue and green traces in

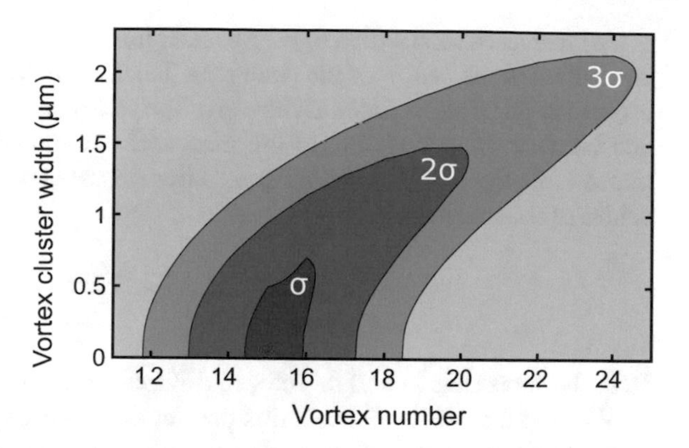

Fig. 5.11 RMS error map for the flat-top distribution decay scenario, computed at $t = 0$. The labelled shaded regions correspond to the one-, two- and three-σ standard deviation confidence intervals. The best fit to experiment at $t = 0$ is provided by 15 vortices near the disk origin in both flat-top and Gaussian scenarios (red circle)

Fig. 5.10b for the uniform and Gaussian models respectively. The calculation of the RMS error for the other decay scenarios follows a similar approach.

Splitting for each mode in scenario 3 (delta-function-like) can be calculated as

$$\Delta f_{\mathrm{pr,i}}^{[3]} = N\,\Delta f_{\mathrm{th,i}}(0), \tag{5.11}$$

Here the optimization is performed over N, the number of circulation quanta around the pedestal. We determine $N(t)$ by minimizing $\Delta f_{\mathrm{RMS\,error}}$ at each of the experimental time steps. We then plot $\Delta f_{\mathrm{RMS\,error}}(t)$, illustrated in orange trace in Fig. 5.10b.

In the vortex dipole model (scenario 4), the spatial probability density $\rho_{\mathrm{free_{N,K}}}$ of the free-orbiting cluster is taken out of a matrix of 19×136 (possible vortex number $\times$ possible radius) possible configurations computed through point-vortex simulations. The delta-function-like distribution approximates the macroscopic circulation pinned to the pedestal, analogous to scenario 3. Then the splitting predicted by the dipole model can be obtained as

$$\Delta f_{\mathrm{pr,i}}^{[4]} = N\left(\Delta f_{\mathrm{th,i}}(0) - \int_0^R \rho_{\mathrm{free_{N,K}}}(r)\,\Delta f_{\mathrm{th,i}}(r)\,r\,\mathrm{d}r\right). \tag{5.12}$$

Since the vortex number in the dipole model is no longer constant due to annihilation events, the optimization at every time step occurs over a two-dimensional space (vortex number and kinetic energy), similar to the initial fitting step of the uniform distribution shown in Fig. 5.11. Having obtained both vortex number and kinetic energy as a function of time, we compute $\Delta f_{\mathrm{RMS\,error}}(t)$, shown in the purple trace in Fig. 5.10b. As we experimentally observe the splitting decaying to zero at long timescales, we make the assumption that the number of circulation quanta pinned around the pedestal matches the number of opposite sign vortices in the free cluster, as shown in Eq. (5.12). This hypothesis is consistent with the pairwise production of vortices by a supercritical superfluid flow, as well as with the requirement for zero net vorticity on a closed surface in three dimensions [50].

All scenarios start with a tightly-packed initial vortex distribution or macroscopic circulation at the centre of the resonator. This is both because we expect the stirring process to generate vortices at the top of the pedestal, where the superflow is fastest, and because the initial ratio of the measured experimental splittings Δf_{exp} closely matches the theoretical splitting Δf_{th} ratios expected from vortices positioned in the centre of the resonator:

$$\frac{\Delta f_{\text{exp},\,m}\,(t=0)}{\Delta f_{\text{exp},m'}\,(t=0)} \simeq \frac{\Delta f_{\text{th},\,m}\,(r=0)}{\Delta f_{\text{th},m'}\,(r=0)} \tag{5.13}$$

Here the subscripts m and m' refer to different sound mode azimuthal orders.

We note that while all scenarios predict a decaying splitting of all sound modes, as observed in the experiments, their relative influence on different sound modes is quite significant. This is striking in Fig. 5.10c(*ii*) and (*iii*), which show the predicted splitting on all five sound modes at the times marked by the dashed grey lines in Fig. 5.10b. We see that both expanding cluster scenarios (flat-top in blue and Gaussian in green) predict similar splitting on all sound modes—on the order of 0.8 kHz in (*ii*) and 0.3 kHz in (*iii*)—as the cluster expands. This is in strong disagreement with the experimental data (black squares) for which the splitting is an increasing function of mode number. This mismatch, visible in the much larger RMS errors for these decay scenarios in Fig. 5.10b, illustrates how the ability to track the frequency splitting of multiple sound modes simultaneously allows us to infer spatial information about the experimental vortex distribution.

We find that the observed splittings are well characterized by a metastable vortex dipole which systematically provides the best agreement with the experimentally measured splittings throughout the entire decay process.

5.8 Coherent Evolution of Vortex Clusters

To determine the dynamics of a metastable vortex distribution from a single continuous measurement, we generate a nonequilibrium vortex cluster by optically initiating supercritical flow. As has been discussed above, the initial metastable distribution is identified as a vortex dipole. The kinetic energy and free-vortex number of the metastable state are shown as a function of time for a single-shot in Fig. 5.12a, b (blue curves). Moreover, Fig. 5.13 demonstrates the temporal evolution of the angular momentum, which is constrained less well than the kinetic energy and free-vortex number. The total kinetic energy of the dipole decays continuously with time over a period of around a minute. This decay time is comparable to previous non-spatially resolved measurements of the decay of a persistent current [54]. In our experiments, the decay is accompanied by a reduction in the number of free vortices, as vortex-vortex interactions and dissipation drive vortices into the centre of the disk where they can annihilate with quanta of circulation of opposite sign pinned to the pedestal. These dynamics are supported by point-vortex simulations (Fig. 5.12c, d), which

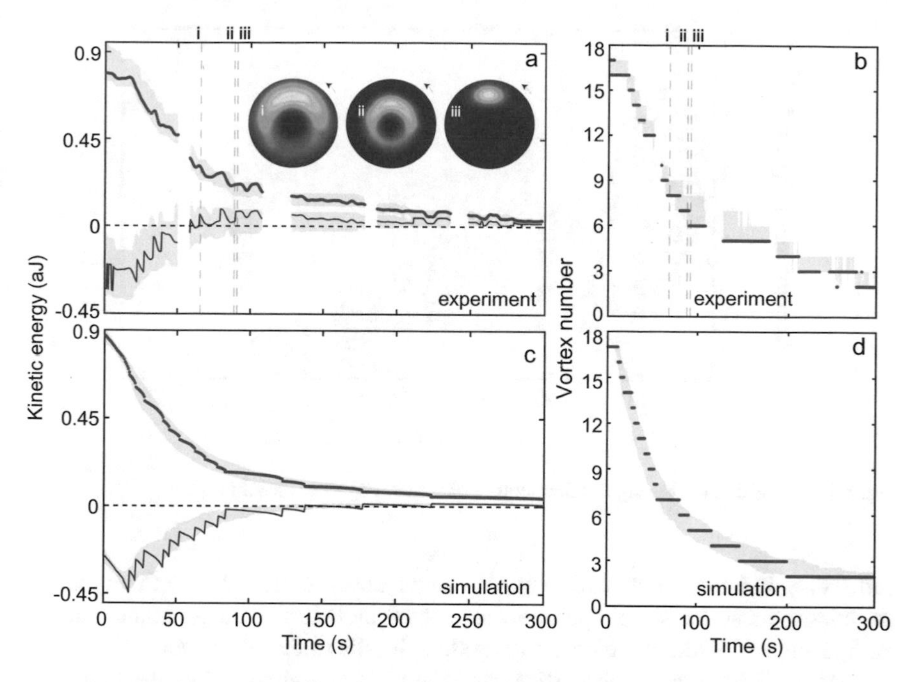

Fig. 5.12 Single-shot evolution of vortex-cluster metastable states. **a** Decay of the total kinetic energy K_{total} (blue curve) and increase of the kinetic energy of the free-vortex cluster K_{free} (red curve). Insets: quasi-equilibrium vortex distributions at times indicated by the vertical dashed lines. The second and third quasi-equilibrium distributions are taken, respectively, just before and just after the 7-to-6 annihilation event. The angular momentum of each distribution is chosen, within the uncertainty window of the fit, to maximize the entropy of the state, and therefore represents the most statistically likely of the experimentally plausible distributions. **b** Experimentally determined decay of the vortex number. Vertical dashed lines correspond to times in (**a**). Note that, while this data displays steps in the vortex number, this is a feature of our analysis which minimizes the root-mean-square uncertainty only over discrete vortex number. While our experiments approach single vortex resolution, the continuous variation of vortex-induced splitting with time precludes direct unambiguous observation of individual steps in the splitting due to the creation or annihilation of vortices [39]. **c** K_{total} decay (blue curve) and K_{free} increase (red curve) extracted from the point-vortex simulation, showing very good agreement with the experimental results in (**a**). **d** Point-vortex simulation of the vortex number decay. In all traces the shaded area corresponds to a one-standard-deviation uncertainty

show good quantitative agreement with only the dimensionless dissipation coefficient γ as a fitting parameter. The dissipation coefficient quantifies the ratio of *coherent* to *dissipative* timescales in the superfluid dynamics, and was found to be $\gamma \sim 2 \times 10^{-6}$ (see Sect. 5.9.3). It has generally been thought that fast dissipative processes would preclude the observation of coherent dynamics in superfluid helium films. However, our experiments show that this is not the case in general, with coherent dynamics dominating by more than five orders-of-magnitude. Indeed, the dissipation coefficient obtained in our work is competitive with the best ultracold atom experiments, which typically achieve $\gamma \sim 6 \times 10^{-4}$ [18, 55]. The agreement between experiment

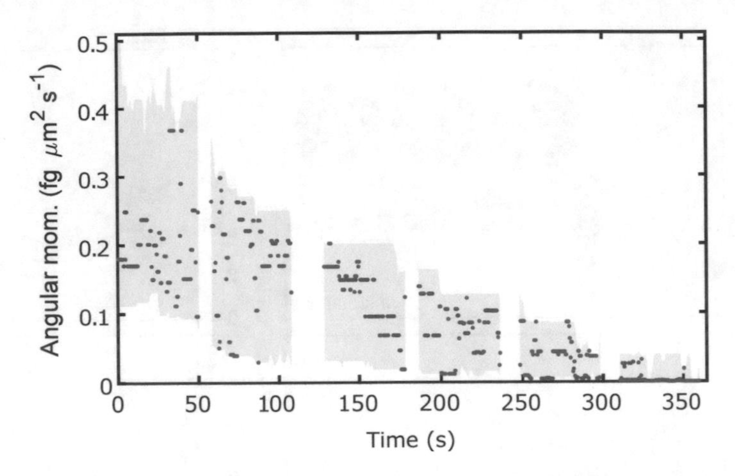

Fig. 5.13 Evolution of the angular momentum of vortex-dipole metastable states

and theory indicates that, within experimental uncertainties, the vortex dynamics are consistent with a simple point-vortex model including local phenomenological dissipation, and without the need to introduce inertia to the vortex cores [24, 25] or an Iordanskii force between vortices and the normal component of the fluid [26]. Thus, we shall now elaborate on the results of the point-vortex simulations and how the point-vortex model is implemented.

5.9 Dissipative Point-Vortex Dynamics

Here we introduce the dissipative point-vortex model (PVM). We employ this model to:

- Calculate the metastable quasi-equilibrium states of the system (Sects. 5.9.2 and 5.9.4).
- Provide a fit to our experimentally measured vortex decay dynamics and extract the dissipation factor γ and diffusion coefficient D (Sect. 5.9.3).

5.9.1 Point-Vortex Model

The motion of quantum vortices of circulation $\kappa_j = s_j\, h/m_{\mathrm{He}}$ (with $s_j \in \mathbb{Z}$ the vortex charge) in a thin superfluid film of density ρ, thickness d, and temperature T can be described by the equation of motion [56, 57]

$$\frac{d\mathbf{r}_i}{dt} = \mathbf{v}_s^i + C(\mathbf{v}_n - \mathbf{v}_s^i) + s_i \left(\frac{Dh\rho d}{m_{\text{He}} k_B T} \right) \hat{\mathbf{z}} \times (\mathbf{v}_n - \mathbf{v}_s^i). \tag{5.14}$$

Here $\mathbf{v}_n$ is the velocity of the normal fluid and $\mathbf{v}_s^i$ is the local superfluid velocity, evaluated at $\mathbf{r}_i$, excluding the self-divergent velocity field of the vortex at $\mathbf{r}_i$ [57]. The parameters C and D are phenomenological mutual friction coefficients, which originate from the scattering of quasiparticles such as phonons, rotons, and ripplons by the vortex cores. These parameters are dependent on temperature, film thickness, and the complex interactions of vortices with defects in the substrate [56, 57]. Note that in the absence of friction, a vortex simply moves with the local superfluid velocity, as in the case of ordinary point-vortex dynamics of an ideal fluid [58]. The diffusion coefficient D is typically $< \hbar/m_{\text{He}}$ when operating well below the superfluid transition, as is the case for our experiments [27]. The constant C ranges between 0 and 1.

In our system the normal fluid is viscously clamped to the surface of the microtoroid [31] and we may therefore assume $\mathbf{v}_n = \mathbf{0}$, giving

$$\frac{d\mathbf{r}_i}{dt} = (1 - C)\mathbf{v}_s^i - s_i\, \gamma\, (\hat{\mathbf{z}} \times \mathbf{v}_s^i), \tag{5.15}$$

where we have defined the dimensionless dissipation coefficient $\gamma = (Dh\rho d/m_{\text{He}} k_B T)$. The superfluid velocity $\mathbf{v}_s$ generated by the vortices is determined by the Green's function of the domain [57]. Here we approximate the surface of the microtoroid as a hard-walled circular disk of radius R. The velocity at vortex i is thus given in terms of the relative vortex positions as

$$\mathbf{v}_s^i = \frac{1}{2\pi} \sum_{j \neq i} \frac{\kappa_j}{r_{ij}^2} \begin{pmatrix} -y_{ij} \\ x_{ij} \end{pmatrix} + \frac{1}{2\pi} \sum_{j} \frac{\bar{\kappa}_j}{\bar{r}_{ij}^2} \begin{pmatrix} -\bar{y}_{ij} \\ \bar{x}_{ij} \end{pmatrix}, \tag{5.16}$$

where $x_{ij} = x_i - x_j$ and, $r_{ij}^2 = x_{ij}^2 + y_{ij}^2$. The barred terms $\bar{x}_{ij} = x_i - \bar{x}_j$, etc. correspond to image vortices with $\bar{\kappa}_j = -\kappa_j$ placed outside the disk at the inverse point $\bar{\mathbf{r}}_i = R^2 \mathbf{r}_i / |\mathbf{r}_i|^2$. These fictitious image vortices enforce the boundary condition $\mathbf{v}_s \cdot \hat{\mathbf{n}}|_{r=R} = 0$, i.e. the flow across the boundary normal $\hat{\mathbf{n}}$ is zero at the boundary $r = R$.

The kinetic energy of the fluid can be expressed in terms of the relative vortex positions. On a disk of radius R, the vortex Hamiltonian for N vortices with core size a_0 is given by [59, 60]

$$\begin{aligned} H = &-\frac{\rho d}{4\pi} \sum_{i<j}^{N,N} \kappa_i \kappa_j \ln \left(\frac{r_{ij}^2}{R a_0} \right) + \frac{\rho d}{4\pi} \sum_{i=1}^{N} \kappa_i^2 \ln \left(\frac{R^2 - r_i^2}{R a_0} \right) \\ &+ \frac{\rho d}{4\pi} \sum_{i<j}^{N,N} \kappa_i \kappa_j \ln \left(\frac{R^4 - 2R^2 \mathbf{r}_i \cdot \mathbf{r}_j + r_i^2 r_j^2}{R^3 a_0} \right), \end{aligned} \tag{5.17}$$

where $r_i = |\mathbf{r}_i|$. In the absence of friction ($\gamma = C = 0$), H generates the vortex dynamics, described in (5.16), from Hamilton's equations as

$$\kappa_i \frac{\mathrm{d}x_i}{\mathrm{d}t} = \frac{\partial H}{\partial y_i}, \qquad \kappa_i \frac{\mathrm{d}y_i}{\mathrm{d}t} = -\frac{\partial H}{\partial x_i}. \tag{5.18}$$

In addition, the angular momentum of the fluid, $L = \int \mathrm{d}^2\mathbf{r}\,(\mathbf{r} \times \mathbf{u})_z$, with velocity field given by $\mathbf{u}$, can be calculated from the position of the point vortices for the case of zero net circulation relevant here as [61]

$$L = -\frac{\rho d R^2}{2} \sum_i \kappa_i r_i^2. \tag{5.19}$$

The parameter C is non-dissipative (it does not influence the relative distances between vortices), and therefore does not contribute to the decay of the superflow [56]. We therefore set $C = 0$. As shall be addressed later in Sect. 5.9.3, we measure the dissipation factor to be on the order of $\gamma \sim 10^{-6}$. As the dimensionless parameters γ and C originate from similar microscopic processes, we expect that they would be of comparable magnitude, as is indeed the case in bulk superfluid helium [62–64]. Since $\gamma \ll 1$, the assumption $(1 - C) \approx 1$ is therefore a reasonable approximation.

We manually add two phenomenological features to the model:

- Pinning of vortices to the pedestal of the microtoroid. In our experiments, we infer a large macroscopic circulation pinned to toroid's pedestal. To model this, we assume rigid pinning, forcing Eq. (5.15) to equal zero for the vortices pinned at the origin.
- Vortex-antivortex annihilation. Vortex-antivortex pairs in a superfluid annihilate once they approach each other within a distance comparable to the core size [65], which is not included in Eq. (5.18). To account for the annihilation of free vortices in the orbiting cluster with quantized circulation of the opposite sign trapped around the pedestal, we remove a free vortex from the simulation when it reaches the radius of the pedestal, $r_p = R/30$, along with one quantum of circulation from the pedestal.

5.9.2 *Metastable Vortex States*

The description of the vortex distributions in our experiments in terms of metastable states relies on a separation of timescales between the dissipative and coherent dynamics. In this case, the system will sample many microstates on a timescale short compared to the time over which the energy, angular momentum and vortex number change, with the average behaviour (i.e. the metastable state) given by the macrostate with the largest entropy [18, 20, 52]. To estimate the characteristic timescale for the

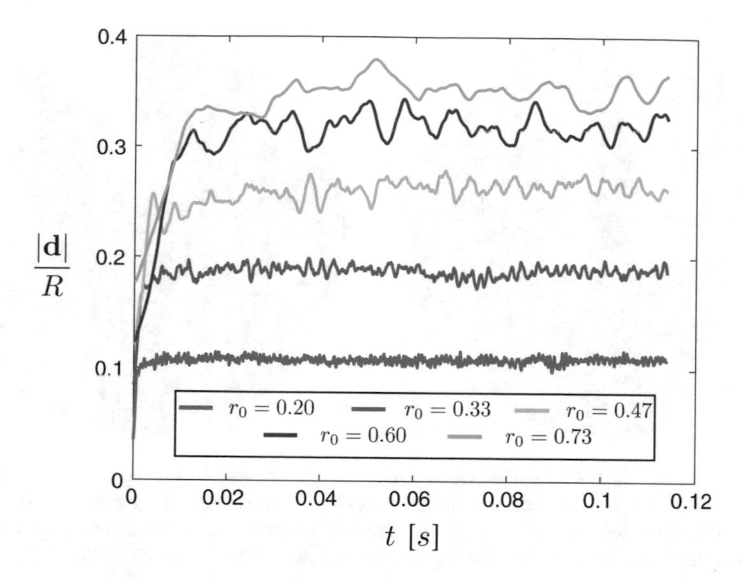

Fig. 5.14 Dipole moment of $N = 17$ vortex cluster, for vortices initialized in a ring formation of varying initial radius r_0. Each curve represents an average of 50 PVM simulation runs. The ring initialization radius r_0 is given here as a fraction of the disk radius R. Figure is reproduced and modified with permission of O. R. Stockdale [66]

coherent dynamics we evolve the conservative ($\gamma = 0$) Hamiltonian dynamics of the cluster and calculate its dipole moment

$$|\mathbf{d}| = \left| \frac{1}{N} \sum_i^N s_i \, \mathbf{r}_i \right|, \tag{5.20}$$

as a function of time. We begin the simulations with a ring of vortices with zero dipole moment. We find that the dipole moment rapidly becomes non-zero as the vortex dipole forms, and asymptotes towards a non-zero value at which point it persists indefinitely. In Fig. 5.14, we plot the dipole moment for a cluster of 17 negative vortices, initialised initially in rings of varying radius r_0 (so as to easily parametrize the cluster), in the presence of a flow generated by a pinned $\kappa = 17\, h/m_{\mathrm{He}}$ vortex at the centre of the disk. Due to the chaotic dynamics of the vortices, the rings become disordered over time, evolving towards a horse-shoe shaped distribution characteristic of a metastable state. Figure 5.14 shows that for all choices of initial radius the dipole moment of each distribution asymptotes to a constant value within a time smaller than $\tau \sim 20$ ms. This value is consistent with the estimation based on the $\tau \sim r_c^2/N\kappa$ formula discussed in Sect. 5.6.

The system therefore does exhibit the required large separation of timescales between the time for mixing (the cluster turnover time) and the timescales for dissipation (minutes) and annihilation events (tens of seconds). As the dynamics are

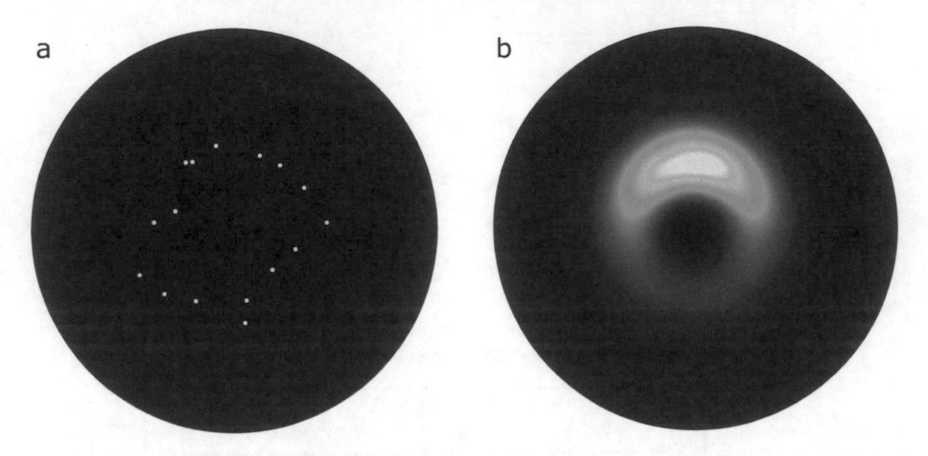

Fig. 5.15 **a** Sample initial ring distribution where $N = 16$. Yellow dots correspond to free vortices in the system, which orbit around the pinned circulation at the centre of the resonator (not shown here). **b** Time-averaged metastable state over the entire simulation. At each time step, the vortex cluster has been rotated such that the dipole moment lies along the y-axis. Figure provided by O. R. Stockdale

expected to be chaotic, on intermediate timescales the system will explore many microstates in a narrow band of kinetic energy and angular momentum values at fixed vortex number. It hence approximately realizes the microcanonical ensemble, with only the energy, angular momentum, and vortex number of the initial distribution defining the metastable state. Using this information, we calculate the metastable states over the entire experimentally-relevant parameter space of (N, L, K). We range the vortex number between $N = 2$ to $N = 19$ (as $N = 1$ corresponds to the trivial case where the vortex simply orbits at a constant radius). For each cluster number, we sample $\sim$270 points in (L, K) space and simulate the conservative point-vortex model ($\gamma = 0$) for $t \simeq 0.6$ s to ensure the cluster has had enough time to sample the metastable state.

We visualize each metastable state by calculating the mean density of vortex positions over the course of the simulation. As the vortices perform many orbits during the simulation, plotting this in a fixed reference frame would produce a doughnut-shaped rotationally invariant density profile. In order to visualize the angular spread of the orbiting cluster, we instead go into a frame rotating with the vortex cluster: at each time step, we rotate the cluster such that its dipole moment lies along the y-axis. An example of this visualisation can be seen in Fig. 5.15b.

5.9.3 Implementation of the Point-Vortex Model

We solve the system of equations in Eq. (5.16) to calculate the motion of vortices with energy damping. We initialize the positions of the 17 negative vortices via a random

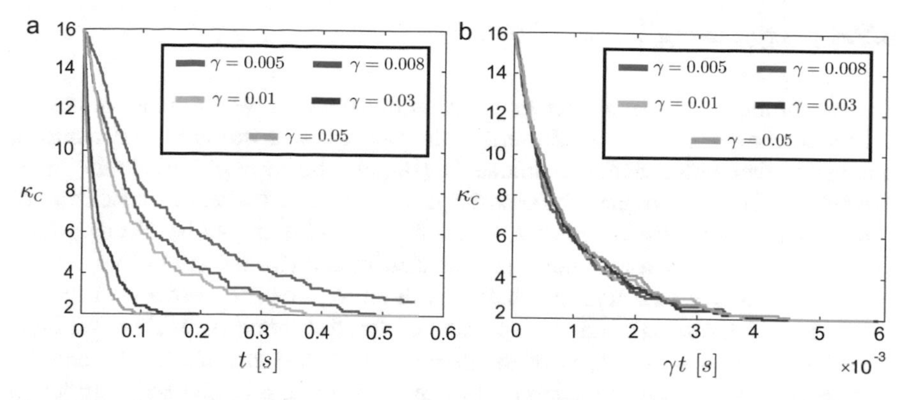

Fig. 5.16 **a** Decay of the charge of the pinned macroscopic circulation (κ_C) for the different dissipation constants given in the legend. Each curve represents the average of 10 runs with different initial conditions constrained by the energy of the cluster. **b** Decay of the charge of the pinned macroscopic circulation (κ_C) as in (**a**), but here the time axis has been scaled by the dimensionless dissipation constant γ. All curves fall on top of each other, highlighting the scaling of the dynamics with γ. Figure is reproduced and modified with permission of O. R. Stockdale [66]

uniform distribution to form a ring, with the condition that the distribution must have a kinetic energy comparable to the initial energy in the experiment (~ 0.9 aJ). We simulate the dynamics of the system thirty separate times with different initial vortex positions, each constrained by initial energy, with dissipation $\gamma = 0.03$. The total kinetic energy of the system is computed using the Hamiltonian in Eq. (5.17). In Fig. 5.12c, d a single simulation result is plotted (solid blue line), with the light-blue shading representing one standard deviation on either side of the mean in the thirty runs.

Since the free-vortex cluster evolves slowly through quasi-equilibrium states, we find that the macroscopic dynamics scale with the dissipation constant γ, as shown in Fig. 5.16. Each curve seen in Fig. 5.16 is an average over 10 runs with different initial conditions. We expect that scaling will improve at lower dissipation rates as the time scale of energy loss will be far smaller than the time scale of coherent dynamics. As such, we choose a value of dissipation that is larger than the experimental one, and scale the results accordingly to best fit the experimental data, rescaling γ in the process. This is necessary due to increasingly large simulation times for decreasing γ. We calculate the motion of vortices in time units of $mR^2/\hbar$. Within our simulations, we effectively run the system for approximately $\sim$50 ms. After scaling the results from the simulations, we find the experiment is consistent with a dissipation factor of $\gamma \sim 2 \times 10^{-6}$.

5.9.4 Expanding Vortex Clusters

For two of the vortex decay models tested in this chapter, we are interested in the evolution of expanding vortex clusters, in the one case with no surface interactions, and in the other with dynamics dominated by hopping between surface pinning sites. We would like to know the characteristic distributions of vortices as a function of time in both cases. As it turns out, two types of evolution can be distinguished: (1) evolution towards a common shape of distribution (for the case of no surface interactions, this is a circular distribution with uniform density, while with surface pinning it is a Gaussian), and (2) expansion of that distribution over time. We hope that the evolution in the shape of the distribution occurs at a faster rate than the expansion, to allow simple models of the expansion process (with fixed distribution shape) to be used in further analysis. This turns out to be the case both with and without pinning. We would note that the rates of both types of evolution are linearly proportional to the dissipation factor, so that this conclusion is independent of the value of the factor.

To study the scenario with no surface pinning, we use the point vortex model to observe the evolution of the density distribution of a vortex cluster initialized with a strongly non-uniform distribution. This simulation is used in Sect. 5.7 to test a possible vortex decay scenario (scenario 1). In Fig. 5.17 we show that the density of a distribution of $N = 1000$ vortices becomes uniform across the disk (despite an initial Gaussian density distribution) on a timescale that is shorter than the timescale of the expansion, i.e. $\tau_{\text{uniform}} < \tau_{\text{expansion}}$. We have performed similar simulations with initial density distributions corresponding to a ring or a flat-top, and observed the same rapid evolution towards uniform density (with respect to the expansion rate). The exact times in Fig. 5.17 are arbitrary and depend upon the magnitude of the dissipation factor chosen. That being said, the interval between each time is constant.

We next simulate the expansion of a vortex cluster under the influence of pinning sites (scenario 2 in Sect. 5.7). As previously, we begin with a Gaussian density distribution of vortices, but phenomenologically include vortex pinning within the point-vortex model. We model vortex pinning by asserting a mean-free-path, ℓ, associated to the likelihood of a single vortex becoming pinned. The probability of a vortex remaining unpinned after travelling a distance Δx is assumed to be exponentially decreasing, so that the probability of a vortex becoming pinned is $P(\text{pin}) = 1 - \exp(-\Delta x/\ell)$.[4] Figure 5.18 shows that the density of a cluster of $N = 100$ vortices remains approximately Gaussian for this situation, but become more spread out over the disk due to energy dissipation.

[4]This simple model assumes an infinite pinning potential, such that once pinned, a vortex is permanently immobilized. Other simulations with a critical pinning velocity below which a vortex gets pinned, which allow for multiple pinning and unpinning events for each vortex, yield qualitatively similar results.

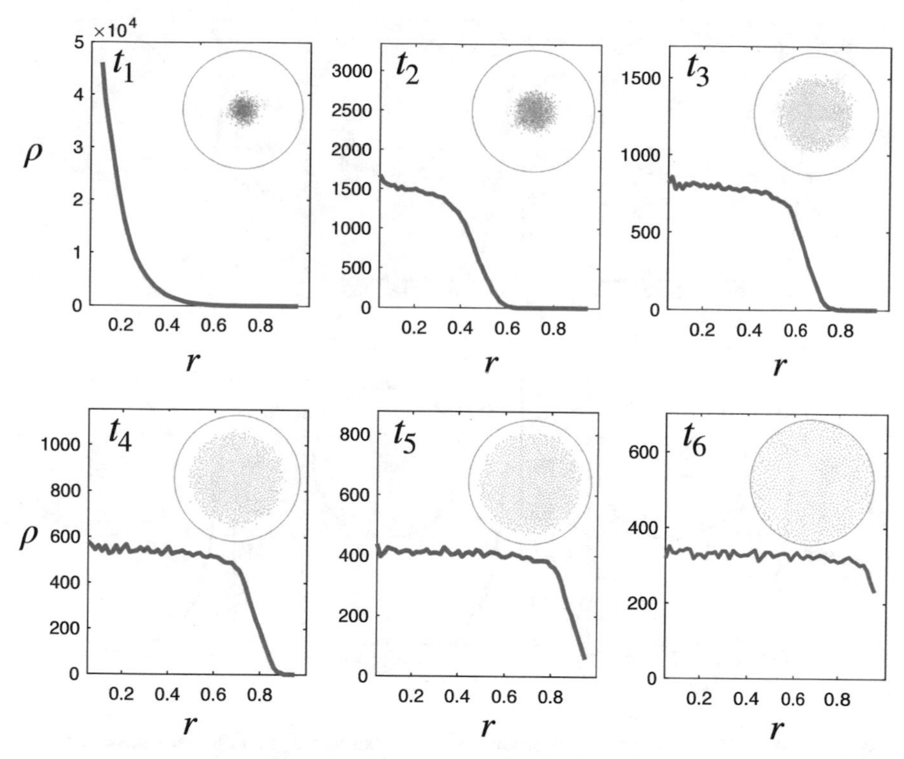

Fig. 5.17 Expansion of a vortex cluster with $N = 1000$ vortices from an initial Gaussian density distribution. Time progression is indicated by the t labels. Even though the initial distribution in non-uniform, vortex dynamics makes the cluster rotate as a rigid body, resulting in an approximately uniform density at later times. This is evident from the insets, which show the physical positions of vortices on the disk tending towards a uniform density as the cluster begins to crystallize. Figure is reproduced and modified with permission of O. R. Stockdale [66]

5.10 Single-Shot Observation of Nonequilibrium Vortex Dynamics

We now move on to describing nonequilibrium vortex dynamics in our system, which we observe in a single shot. The pinning of vortices on the microtoroid pedestal results in a macroscopic circulation, as discussed above. The kinetic energy associated with the free vortices is given by $K_{\text{free}} = K_{\text{total}} - K_{\text{pinned}}$, where $K_{\text{pinned}} = \rho d (N\kappa)^2 \ln\left(R/r_p\right)/4\pi$ is the kinetic energy of the macroscopic circulation alone, with $\rho = 145\,\text{kg/m}^3$ being the density of superfluid helium and $r_p \sim 1\,\mu\text{m}$ the radius of the pedestal. It is shown for both our experimental data and simulations by the red curves in Fig. 5.12a, c. During the first minute of evolution it is negative and, notably, increases with time. The dynamics are characterized by steps up in energy during vortex annihilation events, interspersed with a continuous dissipative decay.

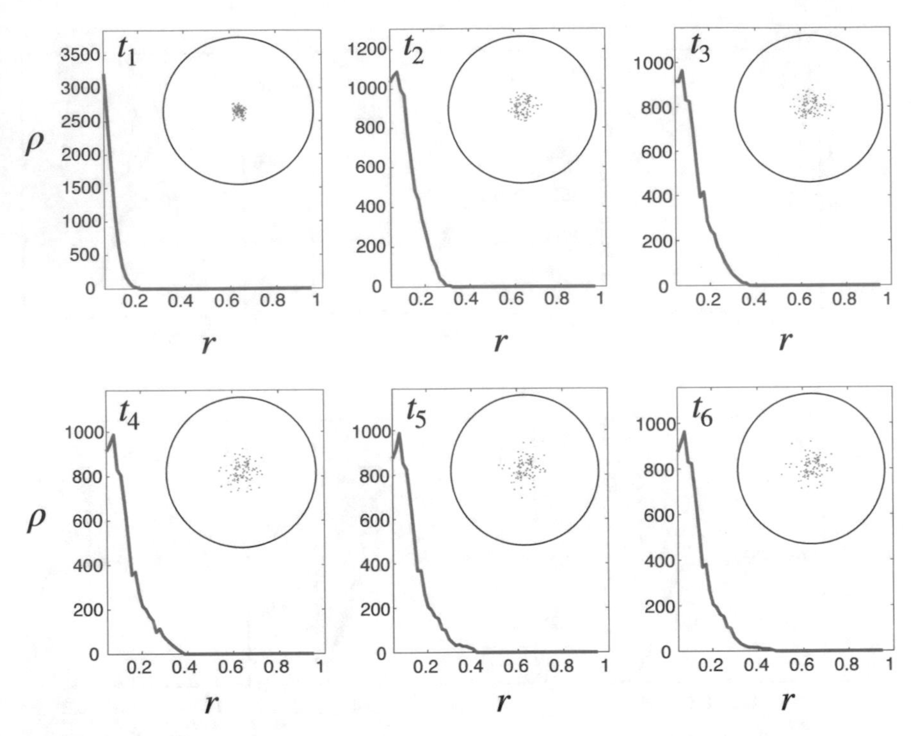

Fig. 5.18 Expansion of a vortex cluster with $N = 100$ vortices from an initial Gaussian density distribution with vortex pinning. Time progression is indicated by the t labels. The pinning qualitatively changes the dynamical evolution of the cluster compared to the no-pinning case (Fig. 5.17). The Gaussian density profile remains approximately Gaussian throughout the simulation, with the standard deviation of the distribution gradually increasing due to damping. Note that by t_4, the cluster is essentially no longer expanding. The choice of the mean-free-path l sets here the maximal cluster width before the vortices freeze out. Figure provided by O. R. Stockdale [66]

The negativity of the free-vortex energy can be understood by considering the interference between the flow fields of a free-vortex and the macroscopic circulation. While the high flow velocity near the core of a free vortex introduces kinetic energy, the vortex flow field also cancels a component of the background flow. For a sufficiently large circulation, this cancellation effect dominates, leading to an overall negative energy cost to introducing the vortex cluster (see Sect. 5.5.3).

The increase in free-vortex energy over time can be explained by considering the process of vortex annihilation in a macroscopic background flow. To annihilate, a free vortex must reach the pedestal, where its contribution to the total kinetic energy is at a minimum. To do this, it gives up kinetic energy to the remaining free vortices. This process of removing low energy vortices has been described as *evaporative heating* [67], in analogy to evaporative cooling of ultracold atomic ensembles [68]. However, while standard evaporative heating can explain a *per-vortex* increase in kinetic energy [19], its effect is to reduce the *net* free-vortex kinetic energy [19]. The physics is modified here by the presence of a macroscopic background flow. The annihilation of a free vortex with a pinned circulation quanta cancels a component of this flow, reducing its kinetic energy while leaving the total kinetic energy essentially unchanged, as can be seen by the lack of discrete steps in the blue curves of Fig. 5.12a, c. We see, therefore, that annihilation events increase the kinetic energy of the free-vortex cluster by drawing energy out of the background flow. This pushes the cluster outwards to a higher separation, as illustrated by the metastable states just before and after an annihilation event in insets ii and iii, respectively, in Fig. 5.12a.

5.11 Vortex Diffusivity and Pinning on Surfaces

As well as allowing the observation of evaporative heating in a single continuous shot, our experiments allow the diffusivity D of vortices to be established in a new regime for two-dimensional superfluid helium. The vortex diffusivity plays an important role in dynamic corrections to the BKT transition [11]. This has motivated substantial research efforts to quantify it both near the BKT transition and outside the regime of BKT superfluidity [27, 34, 54]. However, achieving high signal-to-noise has generally proved challenging [27], while surface pinning is known to have a major influence on measurement outcomes [34, 69]. In contrast, our experiments are not dominated by surface pinning and achieve near-single-vortex resolution. From them, we obtain a value of $D = k_B T \gamma / \rho d \kappa \sim 100\ \mathrm{nm}^2\,\mathrm{s}^{-1}$, where k_B is the Boltzmann constant. This diffusivity is five orders of magnitude below previous measurements where the vortex dynamics is not dominated by pinning [27], and verifies predictions made over thirty years ago that the diffusivity should be exceedingly small outside the BKT regime at low temperatures [27].

The crucial factor which allowed us to obtain such an exceptionally small value of diffusivity is the atomically-smooth surface of our silicon chip. Indeed, the presence of surface roughness on a substrate that is in contact with superfluid helium can significantly limit the ability of vortices to move freely. In thin superfluid films strong pinning sites can freeze out vortex motion altogether [38]. For example, surface roughness is thought to be the primary cause of discrepancy between measurements of the vortex diffusivity around the BKT transition [27]. It is for this reason that studies of vortex interaction, dissipation and diffusion are critically dependent on surface quality.

The effect of surface roughness on vortex motion can be understood by considering the case of a superfluid film residing on a smooth substrate with a defect in the form of a spherical protuberance, as shown in Fig. 5.19a [70, 71]. From this simple picture, one can see that a vortex confined to the defect will be in an energetically favourable state. Indeed, for a constant film thickness d, the vortex located on the defect (center) will have less kinetic energy than the one located off the defect (right) due to its shorter length. This energy difference is associated to a trapping potential, which naturally will grow with increasing defect size. Similarly, from this simple picture

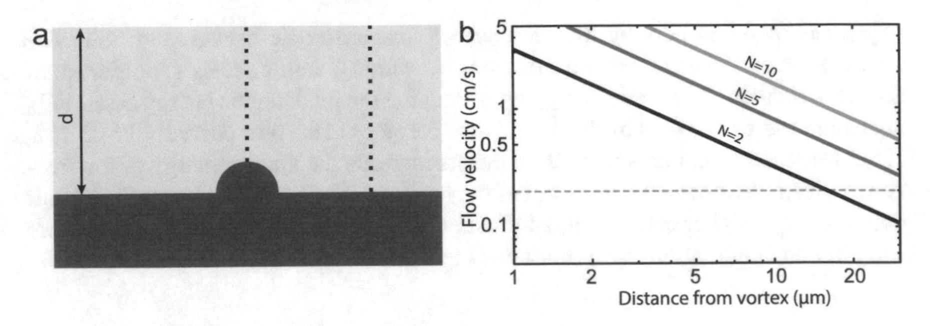

Fig. 5.19 **a** A simplified model of pinning due to a spherical defect in a thin film of superfluid helium of thickness d. The vortex located on the defect (central dashed line) has less kinetic energy than the one off the defect (right), due to its reduced length. **b** Background flow velocity, induced by N circulation quanta trapped around the pedestal, versus radial offset from the disk center. The horizontal line indicates a flow velocity of 0.2 cm/s

one can see that for a given defect size, a thick film will lead to weaker pinning, as the height difference introduced by the defect becomes proportionately smaller.

Depinning occurs when the forces acting upon the vortex due to the background flow exceed the strength of the pinning potential. Clear signatures of depinning in superfluid films of thickness 3 nm have been reported, for example, in Ref. [38]. In that experiment, the frequency splitting of degenerate third-sound modes is slightly reduced when "stirring" the film at flow velocities above 150 cm/s, but below the superfluid critical velocity. The observed change in splitting corresponds to depinning of only a small fraction of the total number vortices, namely 10^2–10^3 compared to 10^5, respectively. This observation suggests that those liberated vortices were attached to weak pinning sites, while the remaining vortices were pinned to much stronger pinning sites, which may be consistent with the comparatively larger surface roughness of their evaporated gold substrate.

In our work we ascribe the lack of any apparent contribution of pinning to the vortex dynamics to the combination of an atomically smooth surface and a relatively thick superfluid film. The dynamics of our system supports a model of a vortex dipole decaying down to a few remaining vortices after a period of minutes. As the vortex dipole decays, the background flow velocity across the disk decreases, as illustrated in Fig. 5.19b. The observed decay is consistent with point-vortex modelling in the absence of pinning sites (see Sect. 5.9), suggesting that pinning does not play a significant role in the vortex dynamics in our system. Towards the end of the experimental run, the background flow induced by two vortices pinned to the pedestal creates flow velocities below $\sim$ 0.2 cm/s over the majority of the disk surface. Indeed, the free-vortex cluster explores these outer-parts of the disk because of vortex annihilations and the evaporative heating effect which push the cluster towards the periphery of the disk. These very low depinning velocities correspond to extremely small surface defects, likely to be of comparable size to the vortex core, consistent with the expected atomic-scale roughness of our commercially purchased silicon wafers.

5.12 Conclusion

The experiments reported in this work were enabled by the greatly enhanced vortex-vortex, vortex-sound, and sound-light interactions provided by microscale confinement. Interactions with strong pinning sites have previously prevented the observation of coherent vortex dynamics in two-dimensional superfluid helium [38]. In our experiments, vortex-vortex interactions dominate due to the increased confinement and atomically smooth surface of the microtoroid. The smoothness of the surface results in a conservative upper bound to the vortex unpinning velocity of 0.2 cm/s, three orders of magnitude lower than in previous experiments [38]. As such, vortex-vortex interactions dominate even for the smallest possible clusters containing only two vortices. Furthermore, a four-order-of-magnitude enhancement in vortex-sound interactions compared to earlier experiments [38, 40] allows resolution approaching the single-vortex level. Together, these capabilities provide a new tool for future study of the rich dynamics of strongly-interacting two-dimensional superfluids on a silicon chip.

The ability to nondestructively track vortex dynamics in a single shot opens the prospect to explore out-of-equilibrium dynamics and stochastic noise-driven processes that are challenging to study with other techniques [16, 72]. It also promises to resolve contentious aspects of vortex dynamics in strongly interacting superfluids, such as dissipation/diffusion models [27, 28], vortex inertia [24, 25] and the Iordanskii force [26]. Furthermore, while the experiments reported here were performed with a relatively small number of vortices, the Ångström-scale of the vortex core in helium-4 will enable the future research on the dynamics of ensembles of thousands of vortices, a regime well outside current capabilities with cold atom and exciton-polariton superfluids [53]. This could allow emergent phenomena in two-dimensional turbulence to be explored, such as the inverse energy cascade [53, 73] and anomalous hydrodynamics [61, 74].

References

1. Mitra D, Brown PT, Guardado-Sanchez E, Kondov SS, Devakul T, Huse DA, Schauß P, Bakr WS (2018) Quantum gas microscopy of an attractive Fermi-Hubbard system. Nat Phys 14(2):173–177
2. Gross C, Bloch I (2017) Quantum simulations with ultracold atoms in optical lattices. Science 357(6355):995–1001
3. King AD, Carrasquilla J, Raymond J, Ozfidan I, Andriyash E, Berkley A, Reis M, Lanting T, Harris R, Altomare F, Boothby K, Bunyk PI, Enderud C, Fréchette A, Hoskinson E, Ladizinsky N, Oh T, Poulin-Lamarre G, Rich C, Sato Y, Smirnov AY, Swenson LJ, Volkmann MH,

Whittaker J, Yao J, Ladizinsky E, Johnson MW, Hilton J, Amin MH (2018) Observation of topological phenomena in a programmable lattice of 1,800 qubits. Nature 560(7719):456
4. Page D, Prakash M, Lattimer JM, Steiner AW (2011) Rapid cooling of the neutron star in Cassiopeia A triggered by neutron superfluidity in dense matter. Phys Rev Lett 106(8):081101
5. Chamel N (2017) Superfluidity and superconductivity in neutron stars. J Astrophys Astron 38(3):43
6. The STAR Collaboration (2017) Global Λ hyperon polarization in nuclear collisions. Nature 548(7665):62–65
7. Kovtun PK, Son DT, Starinets AO (2005) Viscosity in strongly interacting quantum field theories from black hole physics. Phys Rev Lett 94(11):111601
8. Atkins KR (1959) Third and fourth sound in liquid Helium II. Phys Rev 113(4):962–965
9. Mermin ND, Wagner H (1966) Absence of ferromagnetism or antiferromagnetism in one- or two-dimensional isotropic Heisenberg models. Phys Rev Lett 17(22):1133–1136
10. Kosterlitz JM, Thouless DJ (1972) Long range order and metastability in two dimensional solids and superfluids. (Application of dislocation theory). J Phys C: Solid State Phys 5(11):L124
11. Bishop DJ, Reppy JD (1978) Study of the superfluid transition in two-dimensional He 4 films. Phys Rev Lett 40(26):1727
12. Neely TW, Bradley AS, Samson EC, Rooney SJ, Wright EM, Law KJH, Carretero-González R, Kevrekidis PG, Davis MJ, Anderson BP (2013) Characteristics of two-dimensional quantum turbulence in a compressible superfluid. Phys Rev Lett 111(23):235301
13. Freilich DV, Bianchi DM, Kaufman AM, Langin TK, Hall DS (2010) Real-time dynamics of single vortex lines and vortex dipoles in a Bose-Einstein condensate. Science 329(5996):1182–1185
14. Donadello S, Serafini S, Tylutki M, Pitaevskii LP, Dalfovo F, Lamporesi G, Ferrari G (2014) Observation of solitonic vortices in Bose-Einstein condensates. Phys Rev Lett 113(6):065302
15. Amo A, Pigeon S, Sanvitto D, Sala VG, Hivet R, Carusotto I, Pisanello F, Leménager G, Houdré R, Giacobino E, Ciuti C, Bramati A (2011) Polariton superfluids reveal quantum hydrodynamic solitons. Science 332(6034):1167–1170
16. Estrecho E, Gao T, Bobrovska N, Fraser MD, Steger M, Pfeiffer L, West K, Liew TCH, Matuszewski M, Snoke DW, Truscott AG, Ostrovskaya EA (2018) Single-shot condensation of exciton polaritons and the hole burning effect. Nat Commun 9(1):2944
17. Navon N, Gaunt AL, Smith RP, Hadzibabic Z (2015) Critical dynamics of spontaneous symmetry breaking in a homogeneous Bose gas. Science 347(6218):167–170
18. Gauthier G, Reeves MT, Yu X, Bradley AS, Baker MA, Bell TA, Rubinsztein-Dunlop H, Davis MJ, Neely TW (2019) Giant vortex clusters in a two-dimensional quantum fluid. Science 364(6447):1264–1267
19. Johnstone SP, Groszek AJ, Starkey PT, Billington CJ, Simula TP, Helmerson K (2019) Evolution of large-scale flow from turbulence in a two-dimensional superfluid. Science 364(6447):1267–1271
20. Onsager L (1949) Statistical hydrodynamics. Nuovo Cimento, Suppl 6(S2):279–287
21. Makotyn P, Klauss CE, Goldberger DL, Cornell EA, Jin DS (2014) Universal dynamics of a degenerate unitary Bose gas. Nat Phys 10(2):116–119
22. Zwierlein MW, Abo-Shaeer JR, Schirotzek A, Schunck CH, Ketterle W (2005) Vortices and superfluidity in a strongly interacting Fermi gas. Nature 435(7045):1047–1051
23. Anderson PW, Itoh N (1975) Pulsar glitches and restlessness as a hard superfluidity phenomenon. Nature 256(5512):25–27
24. Thouless DJ, Anglin JR (2007) Vortex mass in a superfluid at low frequencies. Phys Rev Lett 99(10):105301
25. Simula T (2018) Vortex mass in a superfluid. Phys Rev A 97(2):023609
26. Sonin EB (1997) Magnus force in superfluids and superconductors. Phys Rev B 55(1):485–501
27. Adams PW, Glaberson WI (1987) Vortex dynamics in superfluid helium films. Phys Rev B 35(10):4633–4652
28. Thompson L, Stamp PCE (2012) Quantum dynamics of a Bose superfluid vortex. Phys Rev Lett 108(18):184501

29. Purdy TP, Grutter KE, Srinivasan K, Taylor JM (2017) Quantum correlations from a room-temperature optomechanical cavity. Science 356(6344):1265–1268
30. Basiri-Esfahani S, Armin A, Forstner S, Bowen WP (2019) Precision ultrasound sensing on a chip. Nat Commun 10(1):132
31. Harris GI, McAuslan DL, Sheridan E, Sachkou Y, Baker C, Bowen WP (2016) Laser cooling and control of excitations in superfluid helium. Nat Phys 12(8):788–793
32. Kashkanova AD, Shkarin AB, Brown CD, Flowers-Jacobs NE, Childress L, Hoch SW, Hohmann L, Ott K, Reichel J, Harris JGE (2017) Superfluid Brillouin optomechanics. Nat Phys 13(1):74–79
33. Noury A, Vergara-Cruz J, Morfin P, Plaçais B, Gordillo MC, Boronat J, Balibar S, Bachtold A (2019) Layering transitions in superfluid helium adsorbed on a carbon nanotube mechanical resonator. arXiv: 1901.09642
34. Gillis KA, Volz SM, Mochel JM (1989) Velocity-dependent dissipation from free vortices and bound vortex pairs below the Kosterlitz-Thouless transition. Phys Rev B 40(10):6684–6694
35. Fonda E, Meichle DP, Ouellette NT, Hormoz S, Lathrop DP (2014) Direct observation of Kelvin waves excited by quantized vortex reconnection. Proc Natl Acad Sci 111:4707–4710
36. Baker CG, Harris GI, McAuslan DL, Sachkou Y, He X, Bowen WP (2016) Theoretical framework for thin film superfluid optomechanics: towards the quantum regime. New J Phys 18(12):123025
37. Bryan GH (1890) On the beats in the vibrations of a revolving cylinder or bell. Proc Camb Philos Soc 7(24):101–111
38. Ellis FM, Li L (1993) Quantum swirling of superfluid helium films. Phys Rev Lett 71(10):1577–1580
39. Forstner S, Sachkou Y, Woolley M, Harris GI, He X, Bowen WP, Baker CG (2019) Modelling of vorticity, sound and their interaction in two-dimensional superfluids. New J Phys 21(5):053029
40. Ellis FM, Luo H (1989) Observation of the persistent-current splitting of a third-sound resonator. Phys Rev B 39(4):2703–2706
41. Hanay MS, Kelber SI, O'Connell CD, Mulvaney P, Sader JE, Roukes ML (2015) Inertial imaging with nanomechanical systems. Nat Nanotechnol 10(4):339–344
42. McAuslan DL, Harris GI, Baker C, Sachkou Y, He X, Sheridan E, Bowen WP (2016) Microphotonic forces from superfluid flow. Phys Rev X 6(2):021012
43. Lee H, Chen T, Li J, Yang KY, Jeon S, Painter O, Vahala KJ (2012) Chemically etched ultrahigh-Q wedge-resonator on a silicon chip. Nat Photonics 6(6):369–373
44. Zhu J, Ozdemir SK, Xiao Y-F, Li L, He L, Chen D-R, Yang L (2010) On-chip single nanoparticle detection and sizing by mode splitting in an ultrahigh-Q microresonator. Nat Photonics 4(1):46–49
45. Baker C, Hease W, Nguyen D-T, Andronico A, Ducci S, Leo G, Favero I (2014) Photoelastic coupling in gallium arsenide optomechanical disk resonators. Opt Express 22(12):14072–14086
46. Tilley DR, Tilley J (1990) Superfluidity and superconductivity. CRC Press
47. Donnelly RJ (1991) Quantized vortices in helium II. Cambridge University Press
48. Donnelly RJ, Fetter AL (1966) Stability of superfluid flow in an annulus. Phys Rev Lett 17(14):747–750
49. Fetter AL (1967) Low-lying superfluid states in a rotating annulus. Phys Rev 153(1):285–296
50. Lamb H (1993) Hydrodynamics. Cambridge University Press
51. Reeves M (2017) Quantum analogues of two-dimensional classical turbulence. PhD thesis, University of Otago
52. Ashbee TL (2014) Dynamics and statistical mechanics of point vortices in bounded domains. PhD thesis, University College London,
53. Reeves MT, Billam TP, Yu X, Bradley AS (2017) Enstrophy cascade in decaying two-dimensional quantum turbulence. Phys Rev Lett 119(18):184502
54. Ekholm DT, Hallock RB (1979) Behavior of persistent currents under conditions of strong decay. Phys Rev Lett 42(7):449–452

55. Bradley AS, Anderson BP (2012) Energy spectra of vortex distributions in two-dimensional quantum turbulence. Phys Rev X 2(4):041001
56. Ambegaokar V, Halperin BI, Nelson DR, Siggia ED (1978) Dissipation in two-dimensional superfluids. Phys Rev Lett 40(12):783–786
57. Ambegaokar V, Halperin BI, Nelson DR, Siggia ED (1980) Dynamics of superfluid films. Phys Rev B 21(5):1806–1826
58. Newton PK (2013) The N-vortex problem: analytical techniques. Springer Science & Business Media
59. Bühler O (2002) Statistical mechanics of strong and weak point vortices in a cylinder. Phys Fluids 14(7):2139–2149
60. Marchand D (2006) Vortex nucleation in a superfluid. PhD Thesis, University of British Columbia
61. Yu X, Bradley AS (2017) Emergent Non-Eulerian hydrodynamics of quantum vortices in two dimensions. Phys Rev Lett 119(18):185301
62. Hänninen R, Baggaley AW (2014) Vortex filament method as a tool for computational visualization of quantum turbulence. Proc Natl Acad Sci 111:4667–4674
63. Barenghi CF, L'vov VS, Roche P-E (2014) Experimental, numerical, and analytical velocity spectra in turbulent quantum fluid. Proc Natl Acad Sci 111:201312548
64. Donnelly RJ, Barenghi CF (1998) The observed properties of liquid helium at the saturated vapor pressure. J Phys Chem Ref Data 27(6):1217–1274
65. Jones CA, Roberts PH (1982) Motions in a Bose condensate. IV. Axisymmetric solitary waves. J Phys A: Math Gen 15(8):2599–2619
66. Stockdale OR (2018) Models of vortex dynamics in thin-film superfluid helium. Honours thesis, The University of Queensland
67. Simula T, Davis MJ, Helmerson K (2014) Emergence of order from turbulence in an isolated planar superfluid. Phys Rev Lett 113(16):165302
68. Anderson MH, Ensher JR, Matthews MR, Wieman CE, Cornell EA (1995) Observation of Bose-Einstein condensation in a dilute atomic vapor. Science 269(5221):198–201
69. Agnolet G, McQueeney DF, Reppy JD (1989) Kosterlitz-Thouless transition in helium films. Phys Rev B 39(13):8934–8958
70. Browne DA, Doniach S (1982) Vortex pinning and the decay of persistent currents in unsaturated superfluid helium films. Phys Rev B 25(1):136–150
71. Schwarz KW (1985) Three-dimensional vortex dynamics in superfluid ^{4}He: line-line and line-boundary interactions. Phys Rev B 31(9):5782–5804
72. Serafini S, Galantucci L, Iseni E, Bienaimé T, Bisset RN, Barenghi CF, Dalfovo F, Lamporesi G, Ferrari G (2017) Vortex reconnections and rebounds in trapped atomic Bose-Einstein condensates. Phys Rev X 7(2):021031
73. Rutgers MA (1998) Forced 2D turbulence: experimental evidence of simultaneous inverse energy and forward enstrophy cascades. Phys Rev Lett 81(11):2244–2247
74. Wiegmann P, Abanov AG (2014) Anomalous hydrodynamics of two-dimensional vortex fluids. Phys Rev Lett 113(3):034501

Chapter 6
Conclusion

The research presented in this thesis advances the understanding of the microscopic dynamics of strongly interacting two-dimensional quantum fluids. Leveraging the state-of-the-art methods of cavity optomechanics, we conducted a comprehensive investigation of some of the fundamental properties of thin films of superfluid helium. The optomechanical system reported in this thesis consists of a high-quality whispering-gallery-mode (WGM) optical microresonator coated with a few-nanometer thick superfluid film. The microscale confinement provided by the optical resonator enables a great enhancement of the interaction strength between the superfluid elementary excitations, such as phonons and quantized vortices. A combination of the small volume of WGM optical modes and the strong optomechanical coupling enhances light-matter interactions at the interface of the microresonator, thus, allowing the microscopic dynamics of the elementary excitations and the interactions between them to be read out optically with unprecedented resolution and precision.

6.1 Summary

Below we provide a summary of the results presented in the thesis.

Chapter 1 introduced the phenomenon of superfluidity and provided a necessary background for the subject of the thesis. This chapter threw a glimpse at optomechanics in Sect. 1.8 and presented the constituents of our optomechanical configuration—WGM optical microresonators and superfluid third-sound modes—in Sect. 1.9 and Sect. 1.10 respectively.

Chapter 2 introduced the paradigm of cavity optomechanics with thin films of superfluid helium. Theoretical aspects of coupling between an optical field, confined within a high-quality whispering-gallery-mode microcavity, and the superfluid

Y. Sachkou, *Probing Two-Dimensional Quantum Fluids with Cavity Optomechanics*, Springer Theses, https://doi.org/10.1007/978-3-030-52766-2_6

mechanical motion were described in this chapter (Sect. 2.2.1). The derivation of an effective mass of the superfluid third-sound modes was presented in Sect. 2.2.2 and showed that thicker films ($\sim$20–30 nm) are more favourable for the quantum regime of optomechanics. Moreover, the chapter introduced the experimental realization of our optomechanical system (Sect. 2.3). Its unprecedented sensitivity and resolution enabled the first demonstration of the capacity to track thermomechanical motion of superfluid helium in real time (Sect. 2.5), i.e. faster than the oscillator's decay timescale. Furthermore, by cooling and heating mechanical modes of thin films of superfluid helium, we showed the ability to control thermal motion of a quantum fluid (Sect. 2.6).

Our optomechanical system equipped us with the capacity to control superfluid flow with light on a chip. This allowed us to demonstrate a new approach to strong microphotonic forcing of mechanical oscillators in cryogenic conditions. Chapter 3 provided both theoretical and experimental descriptions of the light-induced superfluid photoconvective forcing of a mechanical oscillator and showed that the force from the superflow is stronger than its radiation pressure counterpart. We utilized this force to feedback cool a vibrational mode of the microtoroidal resonator down to 137 mK (Sect. 3.3.4).

The capabilities to track phonon excitations of superfluid helium in real time and to control superfluid flow with laser light constructively interfered in our capacity to study vortex-phonon interactions in two-dimensional superfluid helium. Chapter 4 described the theoretical framework for these interactions, which we rigorously quantified. Analytical calculations of the vortex-phonon coupling rate for any arbitrary distribution of vortices within a cylindrical geometry were presented in Sect. 4.2. Furthermore, this chapter introduced a method for the finite-element modelling of the vortex-phonon interactions within an arbitrary domain—not necessarily simply-connected—and for an arbitrary distribution of vortices (Sect. 4.3).

Chapter 5 presented one of the key results of the thesis—the first nondestructive single-shot observation of coherent vortex dynamics in a strongly interacting two-dimensional superfluid. Strong confinement of superfluid helium on the atomically-smooth surface of a silicon chip enabled a great enhancement of coherent interactions between vortices. A combination of the operating temperature much lower than the Berezinskii-Kosterlitz-Thouless phase transition temperature and the atomically-smooth surface of the chip resulted in a vortex diffusivity five orders of magnitude lower than in previous measurements with unpinned vortices in superfluid helium films. In Sect. 5.9 we showed that the observed vortex dynamics is supported by point-vortex simulations. Moreover, in Sect. 5.7 we presented four models of the vortex dynamics feasible in our system and discriminated between them by comparing to the experimental results.

6.2 Directions of Future Research

Our optomechanical system with thin superfluid films is favourable for both cavity optomechanics and quantum fluids. The exploration of strongly interacting superfluids greatly benefits from the ultra-high readout sensitivity and resolution of the system. One of the main goals for the near future is the first observation and tracking of single quantized vortices in two-dimensional superfluid helium. This achievement would open up new prospects for studying quantum turbulence, phase transitions, and dissipation mechanisms in two-dimensional strongly interacting quantum fluids.

On the other hand, the compliant nature of thin superfluid films is highly attractive for applications of optomechanics. Experiments for the near future include, but are not limited to, cooling of a quantum fluid into its quantum-mechanical ground state, demonstration of a superfluid Brillouin laser, control of the phononic eigenmodes with light, etc. Moreover, strong phonon-phonon interactions in thin superfluid films allow a regime of Duffing nonlinearity, which can be exploited for a generation of the quantum superposition state of a macroscopic mechanical oscillator.

Appendix

A.1 Mapping Between Quantities of Superfluid Dynamics, Electrostatics, and Acoustics of an Ideal Gas

Here we demonstrate how quantities and equations of acoustics of an ideal gas can be mapped onto dynamics of superfluids, and how two-dimensional electrostatics can be mapped onto vortex flow fields (Tables A.1 and A.2).

Table A.1 Two-dimensional acoustics, sound dynamics in two-dimensional Bose-Einstein condensates in the Thomas-Fermi limit at zero temperature, and third-sound dynamics on a thin film of superfluid helium

2D-acoustics	Sound in 2D-BEC	Third-sound dynamics
Density perturbation $\alpha(\vec{r}, t)$ [kg/m^2]	Density perturbation $\eta(\vec{r}, t)$ [kg/m^2]	Third-sound amplitude $\eta(\vec{r}, t)$ [m]
Static density ρ_0 [kg/m^2]	Static density ρ_0 [kg/m^2]	Unperturbed film height h_0 [m]
Background flow $u_0(\vec{r})$ [m/s]	Irrotational vortex flow $v_v(\vec{r})$ [m/s]	Irrotational vortex flow $v_v(\vec{r})$ [m/s]
Irrotational flow velocity $\delta\vec{u}(\vec{r}, t)$ [m/s]	Sound flow velocity $\vec{v}_s(\vec{r}, t)$ [m/s]	Third-sound flow velocity $\vec{v}_3(\vec{r}, t)$ [m/s]
Static pressure p_0 [J/m^2]	Atom-atom coupling g_{BEC} [Jm^2]	Linearized VdW coefficient $g = \frac{3\alpha_{\text{vdw}}}{h_0^4}$ [m/s^2]
Speed of sound (acoustics) $c = \sqrt{\gamma RT}$ [m/s]	Bogoliubov sound velocity $c = \sqrt{g_{\text{BEC}} \cdot \rho_0/m^2}$ [m/s]	Speed of sound (thin film) $c_3 = \sqrt{g \cdot h_0}$ [m/s]
Fixed wall boundary $\vec{u} \cdot \vec{n} = 0$	Fixed wall boundary $\vec{v} \cdot \vec{n} = 0$	Free boundary $\vec{v} \cdot \vec{n} = 0$
Fixed pressure boundary $p = p_0$	Fixed density boundary $\eta = 0$	Fixed boundary $\eta = 0$
Continuity equation (acoustics) $\dot{\alpha} = -\rho_0 \vec{\nabla} \cdot \vec{u} - \vec{u}\vec{\nabla}\alpha$	Continuity equation (BEC) $\dot{\eta} = -\rho_0 \vec{\nabla} \cdot \vec{v} - \vec{v}\vec{\nabla}\eta$	Continuity equation (thin film) $\dot{\eta} = -h_0 \vec{\nabla} \cdot \vec{v} - \vec{v}\vec{\nabla}\eta$
Linearized Euler (acoustics) $\dot{\vec{u}} + (\vec{u} \cdot \vec{\nabla})\vec{u} = -\frac{\gamma RT}{\rho_0} \vec{\nabla}\alpha$	Linearized Euler (BEC) $\dot{\vec{v}} + (\vec{v} \cdot \vec{\nabla})\vec{v} = -\frac{1}{m}\vec{\nabla}(U + \frac{\eta}{m} g_{\text{BEC}})$	Linearized Euler (thin film) $\dot{\vec{v}} + (\vec{v} \cdot \vec{\nabla})\vec{v} = -g\vec{\nabla}\eta$

Y. Sachkou, *Probing Two-Dimensional Quantum Fluids with Cavity Optomechanics*, Springer Theses, https://doi.org/10.1007/978-3-030-52766-2

Table A.2 Mapping between electrostatics and vortex flow field. Units and equations are presented in two dimensions

2D-electrostatics	Vortices
Electric displacement field $\vec{D}(\vec{r})$ [C/m^2]	Velocity field $\vec{v}_v(\vec{r})$ [m/s]
Electric line charge Q [C/m]	Circulation quantum κ [m^2/s]
Gauss's law $\oint \vec{D} \cdot d\vec{n} = Q$	Vortex flow equation $\oint \vec{v}_v \cdot d\vec{l} = \kappa$
Perfect electric conductor (*ground*) $\vec{D} \times \vec{n} = 0$	Tangential flow boundary $\vec{v}_v \cdot \vec{n} = 0$

Curriculum Vitae

YAUHEN (EUGENE) SACHKOU
The University of Queensland, Australia
yauhen.sachkou@uq.net.au

EDUCATION

2015–2019 PhD in Physics, University of Queensland
2012–2014 European Master of Science in Photonics (with Great Distinction),
2007–2012 Belarusian State University (summa cum laude diploma)

RESEARCH EXPERIENCE

Present **University of Queensland**
Research on superconducting quantum circuits (Superconducting Quantum Devices Laboratory)

2015–2019 **University of Queensland**
Research on superfluid optomechanics (Queensland Quantum Optics Laboratory)

2014 **University of St Andrews**
Research on quantum correlations in light (Quantum Optics Laboratory)

2013 **University of St Andrews**
Internship (Quantum Optics Laboratory)

Experience in other research areas
Semiconductor lasers, metamaterials, plasmonics, graphene photonics

Y. Sachkou, *Probing Two-Dimensional Quantum Fluids with Cavity Optomechanics*, Springer Theses, https://doi.org/10.1007/978-3-030-52766-2

DISTINCTIONS AND AWARDS

2020 Springer Thesis Award for an "Outstanding Ph.D. Research"
2020 Australian Academy of Science Fellowship to the Lindau Nobel Laureate Meetings
2020 Australian delegate to the 70th Lindau Nobel Laureate Meeting
2016 University of Queensland International Scholarship
2012 European Master of Science in Photonics Excellence Grant
Before 2012 Fellowship of the Ministry of Education of Belarus; Scholarship of the Belarusian State University

LEADERSHIP

2016–2018 University of Queensland Physics Colloquium coordinator
2017 Co-organiser of the OSA IONS KOALA international conference
2016–2018 University of Queensland OSA Student Chapter vice-president
2010–2012 Vice-chair of the Belarusian State University IEEE branch
2010–2012 Founder and chair of the Students Physics Society at the Belarusian State University

LIST OF PUBLICATIONS

- **Y. P. Sachkou**, C. G. Baker, G. I. Harris, O. R. Stockdale, S. Forstner, M. T. Reeves, X. He, D. L. McAuslan, A. S. Bradley, M. J. Davis, W. P. Bowen. Coherent vortex dynamics in a strongly interacting superfluid on a silicon chip. *Science* 366, 1480–1485, 2019.
- S. Forstner, **Y. Sachkou**, M. Woolley, G. I. Harris, X. He, W. P. Bowen, C. G. Baker. Modelling of vorticity, sound and their interaction in two-dimensional superfluids. *New Journal of Physics* 21, 053029, 2019.
- X. He, G. I. Harris, C. G. Baker, A. Sawadsky, Y. L. Sfendla, **Y. P. Sachkou**, S. Forstner, W. P. Bowen. Strong optical coupling through superfluid Brillouin lasing. *Nature Physics* 16, 417–421, 2020.
- G. I. Harris, D. L. McAuslan, E. Sheridan, **Y. Sachkou**, C. Baker, W. P. Bowen. Laser cooling and control of excitations in superfluid helium. *Nature Physics* 12, 788–793, 2016.
- D. L. McAuslan, G. I. Harris, C. Baker, **Y. Sachkou**, X. He, E. Sheridan, W. P. Bowen. Microphotonic forces from superfluid flow. *Physical Review X* 6, 021012, 2016.
- C. G. Baker, G. I. Harris, D. L. McAuslan, **Y. Sachkou**, X. He, W. P. Bowen. Theoretical framework for thin film superfluid optomechanics: towards the quantum regime. *New Journal of Physics* 18, 123025, 2016.

INVITED CONFERENCE TALKS

- Vortex dynamics and evaporative heating in a strongly interacting superfluid, **Y. Sachkou**, C. Baker, G. Harris etc. The EQUS Annual Workshop, December 2018, Perth, WA, Australia.
- Probing the dynamics of 2D superfluids with cavity optomechanics, **Y. Sachkou**, C. Baker, X. He, S. Forstner, R. Kalra, and W. Bowen. Quantum Optics and Quantum Information, November 2017, Minsk, Belarus.
- Optomechanical measurements of quantum fluid dynamics, **Y. Sachkou**, C. Baker, X. He, S. Forstner, R. Kalra, and W. Bowen. Optomechanics Incubator, December 2016, Brisbane, Australia.
- Resonators in liquids and liquids as resonators (Discussion Leader talk), **Y. Sachkou**. Gordon Research Seminar "Mechanical Systems in the Quantum Regime", March 2016, Ventura, CA, USA. Discussion Leader for the "Resonators in Liquids and Liquids as Resonators" session.

Printed in the United States
by Baker & Taylor Publisher Services